GEOTECHNICAL SPECIAL PUBLICATION NO. 107

SOIL DYNAMICS AND LIQUEFACTION 2000

PROCEEDINGS OF SESSIONS OF GEO-DENVER 2000

SPONSORED BY
The Geo-Institute of the American Society of Civil Engineers

August 5-8, 2000
Denver, Colorado

EDITED BY
Ronald Y.S. Pak
Jerry Yamamura

D0108960

 American Society
of Civil Engineers
1801 ALEXANDER BELL DRIVE
RESTON, VIRGINIA 20191–4400

Abstract: This Geotechnical Special Publication contains papers on contemporary subjects of active research and practical interest in the field of soil dynamics and geotechnical earthquake engineering. Topics include methods for liquefaction assessment, constitutive modeling, cyclic shear strength of soils, dynamic soil–foundation interaction, dam engineering, soil improvement, in situ testing, ground motion zonation, and probabilistic methods.

Library of Congress Cataloging-in-Publication Data

Geo-Denver 2000 (2000 : Denver, Colo.)
   Soil dynamics and liquefaction 2000 : proceedings of sessions of Geo-Denver 2000 : August 5-8, 2000, Denver, Colorado / sponsored by The Geo-Institute of the American Society of Civil Engineers ; edited by Ronald Y.S. Pak, Jerry Yamamura.
      p. cm. – (Geotechnical special publication ; no. 107)
   Includes bibliographical references and index.
   ISBN 0-7844-0520-4
   1. Soil dynamics—Congresses. 2. Soil liquefaction—Congresses. 3. Soil structure interaction—Congresses. I. Pak, Ronald Y. S. II. Yamamura, Jerry. III. American Society of Civil Engineers. Geo-Institute. IV. Title. V. Series.

TA711.A1 G46 2000a
624.1'5136--dc21   00-042133

Any statements expressed in these materials are those of the individual authors and do not necessarily represent the views of ASCE, which takes no responsibility for any statement made herein. No reference made in this publication to any specific method, product, process or service constitutes or implies an endorsement, recommendation, or warranty thereof by ASCE. The materials are for general information only and do not represent a standard of ASCE, nor are they intended as a reference in purchase specifications, contracts, regulations, statutes, or any other legal document. ASCE makes no representation or warranty of any kind, whether express or implied, concerning the accuracy, completeness, suitability, or utility of any information, apparatus, product, or process discussed in this publication, and assumes no liability therefore. This information should not be used without first securing competent advice with respect to its suitability for any general or specific application. Anyone utilizing this information assumes all liability arising from such use, including but not limited to infringement of any patent or patents.
Photocopies: Authorization to photocopy material for internal or personal use under circumstances not falling within the fair use provisions of the Copyright Act is granted by ASCE to libraries and other users registered with the Copyright Clearance Center (CCC) Transactional Reporting Service, provided that the base fee of $8.00 per article plus $.50 per page is paid directly to CCC, 222 Rosewood Drive, Danvers, MA 01923. The identification for ASCE Books is 0-7844-0520-4/00/ $8.00 + $.50 per page. Requests for special permission or bulk copying should be addressed to Permissions & Copyright Dept., ASCE.

Copyright © 2000 by the American Society of Civil Engineers, All Rights Reserved.
Library of Congress Catalog Card No: 00-042133
ISBN 0-7844-0520-4
Manufactured in the United States of America.

Geotechnical Special Publications

Foreword

At the start of our new millennium, one can look back and celebrate the considerable progress that we have made in the field of soil dynamics and geotechnical earthquake engineering in the last century. As in any field of engineering, there are techniques and methods in soil dynamics that were developed to serve the immediate needs of the state of the practice, and there were studies and investigations that are critical to the scientific growth of our discipline. Without the help of the former, we will be hard pressed to fulfill our responsibilities as engineers to deal with the urgent needs of our society. Without the direction of the latter, we will run the risk of straying away from the truth and depriving our discipline of parallel progress with science and technology. As a modest platform to see where we are now in terms of ideas and concepts, this GSP contains papers on a variety of contemporary subjects of active research and practical interest. The topics include modern methods of liquefaction assessment, constitutive modeling, cyclic shear strength, dynamic soil-structure interaction, dam engineering, soil improvement, in-situ testing, and probabilistic methods. From these communications, one will see for instance how combinations of modern computational methods and experimental research can help to advance the understanding of both fundamental and applied problems. As should also be apparent from the ensemble, however, there are still a lot to be learnt, misconceptions to be dispelled, and rationality to be gained in our approaches to many soil dynamics and liquefaction problems. We like to thank the authors for their valuable contributions and hope the readers will find useful inspirations from their research and observations. All the papers have undergone the standard ASCE Geo-Institute's special publication review process and are eligible for discussions and ASCE awards.

Ronald Y.S. Pak
Professor
University of Colorado, Boulder

Jerry Yamamura
Assistant Professor
University of Delaware

Contents

Theoretical Approach to Sand Liquefaction

N. D. Cristescu[1]

Abstract

The paper presents a non associated elasto/viscoplastic constitutive equation for saturated sand. This constitutive equation assumes: instantaneous elastic response, yield surfaces in the sense of viscoplasticity (i.e., after each reloading following with delay in time the loading path), it describes both compressibility and dilatancy and the constitutive equation is non associated (i.e., the viscoplastic potential does not coincide with the yield function). It is shown that in this case a strip in the constitutive domain exists where short-term repeated loading/unloading pulse can produce loosing of stability, i.e., liquefaction.

Introduction.

Very many papers have been devoted to sand liquefaction. Most of these paper are case studies or laboratory tests, but there are also theoretical approaches in which various models have been used (see, for instance, Han and Vardoulakis [1991], Vaid and Sasitharan [1992], Anandarajah [1994], Nemat-Nasser and Shokooh [1980], Veyera and Charlie [1990], besides many others). An excellent review paper presenting laboratory tests and case studies is due to Ishihara [1993]. All main features of sand liquefaction are presented with great accuracy.

In this paper we are presenting a theoretical approach to sand liquefaction. One starts by assuming that, mainly in dynamic problems, saturated sand satisfies an elasto/viscoplastic nonassociated constitutive equation. Thus, the instantaneous response is assumed elastic, in the sense that the two extended body elastic waves (longitudinal and transverse) can propagate in sand. The yield stress can be assumed to be essentially zero. The yield surfaces are surfaces of equal stored (or partially released) energy. The yield surfaces in viscoplasticity are not following

---

[1] Graduate Research Professor; Dept. of Aerospace Engineering, Mechanics & Engineering Science, University of Florida, P.O.Box 116250, Gainesville, FL. 32611-6250

1

instantaneously the stress path, but with time delay. Thus, an "instantaneous" loading followed immediately by an 'instantaneous" unloading will not change at all the yield surface. Therefore, a dynamic loading/unloading cycle performed in a very short time interval, will influence very little the yield surface. The constitutive equation formulated for saturated sand is describing both compressibility and dilatancy. Also, it has been found (see literature mentioned by Cristescu and Hunsche [1998]) that only nonassociated constitutive equations can describe accurately the mechanical behavior of saturated sand: thus the yield function does not coincide with the viscoplastic potential. These are the arguments used in the theoretical approach of sand liquefaction:

- Instantaneous elastic response to "instantaneous" loading or unloading;
- the constitutive equation describes both compressibility and dilatancy:
- the constitutive equation is viscoplasticity, i.e., all irreversible deformation is viscoplastic:
- the constitutive equation is nonassociated.

The present analysis was published as a short paragraph in a paper devoted to constitutive equation for sand (Cristescu [1991]) (and sent to the editor of the journal in 1989 already). Since this journal (Int. J. Plasticity) was not dedicated to geomechanics problems, this analysis remained unnoticed by the specialists in the field.

Constitutive equation

The constitutive equation used is of the same kind as used for rocks (Cristescu [1989, 1999], Cristescu and Hunsche [1998]) or for particulate materials (Cristescu [1991, 1996], Cazacu, Jin and Cristescu [1997], Cristescu, Cazacu and Jin [1997], Jin and Cristescu [1998]):

$$D = \frac{\dot{T}}{2G} + \left(\frac{1}{3K} - \frac{1}{2G}\right)\dot{\sigma}\,\mathbf{1} + k\left\langle 1 - \frac{W(t)}{H(T)}\right\rangle \frac{\partial F(T)}{\partial T} \tag{1}$$

where **D** is the rate of deformation tensor, **T** - the Cauchy stress tensor, σ - the mean stress,

$$W(T) = \int_0^T \sigma(t)\dot{\varepsilon}_v^I(t)\,dt + \int_0^T T'(t):D'^I(t)\,dt \tag{2}$$

is the irreversible stress work per unit volume, "prime " stands for deviator and ε_v^I is the irreversible volumetric strain. Further in (1)   H(**T**) is the yield function depending on stress invariants and

$$H(\mathbf{T}(t)) = W(t) \qquad (3)$$

is the equation of stabilization boundary (stabilization of transient creep and/or stress relaxation). Finally $F(\mathbf{T})$ is the viscoplastic potential depending also on stress invariants, and k a viscosity parameter. The irreversible volumetric strain rate is obtained from

$$\dot{\varepsilon}_v = k \left\langle 1 - \frac{W(t)}{H(\mathbf{T})} \right\rangle \frac{\partial F}{\partial \sigma} \qquad (4)$$

where the derivative of F is with respect to the mean stress σ. Thus, dilatancy takes place there where $\partial F / \partial \sigma < 0$, while compressibility there where $\partial F / \partial \sigma > 0$. The compressibility/dilatancy boundary is defined by $\partial F / \partial \sigma = 0$.

The first two right-hand side terms in (1) describe the "instantaneous" response, while the last term the "time-effects", or non-instantaneous viscoplastic behavior.

All the above mentioned functions or parameters are determined from a few tests performed in triaxial apparatus. The procedure follows these steps:

First are determined in triaxial tests the **elastic parameters** from short loading/unloading cycles performed at various stress levels, after keeping the stress constant for short period of time (15 – 30 minutes) (see Cristescu [1989], Nawrocki et al. [1999]). The reason for this procedure is that if the unloading is performed just after a loading, a significant hysteresis loop is generally observed. If, after performing the loading up to a desired stress level one keeps this stress constant for a short time interval, this hysteresis loop is disappearing during a small unloading/reloading cycle, i.e. the rheological properties are "separated" quite well from unloading. Also, by this procedure the "quasistatic" determined elastic parameters are quite close to the "dynamic" ones.

Afterwards is found from tests the **yield function** $H(\mathbf{T})$ without any a priori assumption, by estimating the deformation energy (i.e., $W(t)$). The first right hand term in (2) is the volumetric stress work either stored (in the compressibility domain) or released (in the dilatancy domain). First, the volumetric stress work per unit volume is determined in hydrostatic tests (when the first right-hand side term in (2) is the only one nonzero. Afterwards, both right-hand side terms from (2) are determined in the deviatoric stage of the triaxial tests. Thus is found the yield function involved in (3).

The last step is to find the **viscoplastic potential** by determining the derivatives of F with respect to the stress invariants. That can be done after determining the yield functions $H(\mathbf{T})$ and knowing from tests the orientation of the rate of deformation components along the triaxial loading paths. Why the viscoplastic potential is distinct from the yield function? The viscoplastic potential is defining the orientation of the irreversible strain rate tensor. Therefore according to (4), this function is making precise for what stress state the geomaterial is in a compressible state, for what stress state at is in a dilatant state and finally for what stress state the geomaterial passes from compressibility to dilatancy. The compressibility/dilatancy boundary is therefore defined by $\partial F / \partial \sigma = 0$ (see (4)). If the constitutive law would be associated the equation of this boundary would be just $\partial H / \partial \sigma = 0$. After determining the boundary $\partial H / \partial \sigma = 0$ on can compare it with the compressibility/dilatancy boundary determined from experiment. For sand, as for most geomaterials, the boundary $\partial H / \partial \sigma = 0$ is quite far from the compressibility/dilatancy boundary (see Fig.1). That is why if one wishes to determine accurately the irreversible behaviour of the volume one has to determine in some other way the orientation of the irreversible strain rate tensor, as for instance, by a viscoplastic potential or in a simpler way by a strain-rate orientation tensor (Cristescu and Hunsche [1998]).

In order to determine $F(\mathbf{T})$, one starts from the determination of the equations of the C/D boundary from tests; for saturated sand it can be approximated by $-\bar{\sigma} + 2 f \sigma = 0$ with $f = 0.562$. Then $(2f + \alpha)\sigma - \left(1 - \dfrac{\alpha}{3}\right)\bar{\sigma} = 0$ is the equation of the short-term failure surface, with $\alpha = 1.34$. The determination of F starts from the determination of the derivative $\partial F / \partial \sigma$. For sand one can use the formula

$$\frac{\partial F}{\partial \sigma} = h_1 \frac{\left(-\bar{\sigma} + 2 f \sigma\right)\sqrt{\sigma}}{\left(2f + \alpha\right)\sigma - \left(1 + \dfrac{\alpha}{3}\right)\bar{\sigma}} \tag{5}$$

where the term $\sqrt{\sigma}$ comes from the matching of the data on sand compressibility in hydrostatic tests. Thus (5) defines quite accurately the passage from compressibility to dilatancy and vice versa, since the equation of the C/D boundary as determined from tests is incorporated in the function F.

For further details of the procedure used to determine the function F see Cristescu [1991] or Cristescu and Hunsche [1998].

For saturated fine silica sand the data by Lade *et al.* [1987] have been used. The yield surfaces H = const. and viscoplastic potential surfaces F = const., as determined from tests, using the procedure described shortly above, and without any a priori assumption concerning their shapes, are shown in Fig.1 (Cristescu [1991]). It is obvious that the two families of surfaces: the yield surfaces H = const. (dotted lines) and viscoplastic potential surfaces F = const. (interrupted lines) are quite distinct. The full line is the short-term failure surface. In Fig.1 $\bar{\sigma}$ is the equivalent stress $\bar{\sigma} = \sqrt{3 \, II_{T'}}$.

An important feature of the constitutive equation (1) and which will be involved in the theoretical approach to liquefaction is the following. Let us assume

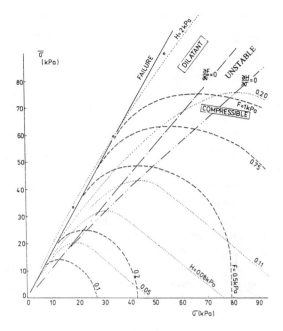

Fig.1 Yield surface $H = $ const, viscoplastic potential surfaces $F = $ const, failure line (full line), compressibility/dilatancy boundary $\partial F / \partial \sigma = 0$ and loosing of stability boundary $\partial H / \partial \sigma = 0$.

that the actual stress states is $\mathbf{T}^P$ and the corresponding value of W in an equilibrium state (i.e., satisfying (3)) is W^P. If at time t_o the stress state is suddenly changed to $\mathbf{T}(t_o)$ in a loading process (i.e. $H(\mathbf{T}(t_o)) > H(\mathbf{T}^P)$) and is afterwards kept constant, $W(t)$ is varying according to (Cristescu [1989] § 8.4)

$$\frac{W(t)}{H(\mathbf{T}(t_o))} = 1 + \left(\frac{W^P}{H(\mathbf{T}(t_o))} - 1 \right) exp \left[- \frac{k}{H} \frac{\partial F}{\partial \mathbf{T}} : \mathbf{T}(t - t_o) \right]. \tag{6}$$

Thus the yield surface is following the sudden loading with a significant delay. If, however, the fast loading is followed immediately by a fast unloading (vibratory loading), the change in the yield surface according to (6), is extremely small. This depends on the "relaxation time" which, in our case, is depending on stress state (the reciprocal of $\dfrac{k}{H}\dfrac{\partial F}{\partial T}:T$ computed for $T(t_0)$).

A possible theoretical approach to sand liquefaction.

First we would like to mention again that the compressibility/ dilatancy boundary $\partial F/\partial \sigma = 0$ is quite precisely determined, since this boundary, as determined from tests, is incorporated exactly as it is, in the expression of the function F (see (5)). The line (generally not a straight line) $\partial F/\partial \sigma = 0$ is passing by the maxima of the lines F = const. (see Fig.1). At the right of this curve $\partial F/\partial \sigma > 0$ and the sand is irreversible compressible, i.e., $\dot{\varepsilon}_v^I > 0$, (see (4)). At the left of this curve $\partial F/\partial \sigma < 0$ and the sand is irreversible dilatant $\dot{\varepsilon}_v^I < 0$.

The line $\partial H/\partial \sigma = 0$ is not so precisely determined, since the function $H(T)$ is determined following a long procedure to estimate the deformation energy along the triaxial loading paths. After determining the surfaces H = const the line $\partial H/\partial \sigma = 0$ is the one uniting the maxima of the curves H = const. shown in Fig.1. If the constitutive equation would be associated then the line $\partial H/\partial \sigma = 0$

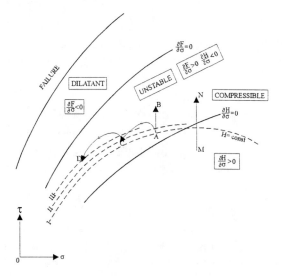

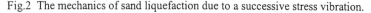

Fig.2 The mechanics of sand liquefaction due to a successive stress vibration.

would be just the compressibility/dilatancy boundary. But the constitutive equation equation is not associated. This line $\partial H/\partial \sigma = 0$ has another meaning which will be revealed below. It will be shown that in the domain where both inequalities

$$\frac{\partial F}{\partial \sigma} > 0 \quad , \frac{\partial H}{\partial \sigma} < 0 \qquad (7)$$

are simultaneous satisfied, a saturated sand may become unstable if subjected to several dynamic loading/unloading pulses (Cristescu [1991]). That is shown schematically in Fig.2 . This figure shows a portion of the strip bounded between the two lines $\partial F/\partial \sigma = 0$ and $\partial H/\partial \sigma = 0$ shown in Fig.1. This is a strip of the "compressibility" domain $(\partial F/\partial \sigma > 0)$ where however $\partial H/\partial \sigma < 0$, i.e., the slopes of the H = const. curves shown in Fig.1 have a normal which, if projected on the σ axis is oriented towards the negative direction.

Let us assume that the stress state existing at a certain location in the sand mass (under a building, say, etc.) is represented by point A belonging to the strip (7) (Fig.2) and that this is an equilibrium state. If a dynamic loading/unloading stress pulse is superposed the sand may not be able to carry any more this load. For simplicity let us assume that the stress vibration is a shearing loading increasing the octahedral shear stress τ only, for instance along AB. If the sand is saturated such variation (i.e. only of τ) is not possible. When τ begins to increase, since this loading path belongs entirely to the compressibility domain, the solid skeleton is subjected to a dynamic compressibility. That will produce an increase of the pore pressure, which in turn will reduce the mean stress in the skeleton. Thus each such pulse will produce a decrease of the mean stress in the skeleton. On the other hand since the pulse is of short duration the initial yield surface (position I in the Figure 1) will move (change) very little, during this pulse, to the position II, say. Thus the real loading stress trajectory is AC instead of AB. The whole segment AC means loading. The final stress state may be located even under the yield surface shown as II. The second pulse, if it is reaching points outside the yield surface II will produce an additional decrease of the mean stress in the skeleton. Due to the peculiar shape of the yield surfaces H = const. in this strip (7), each such pulse will produce a decrease of the mean stress and, maybe, that of τ needed to produce yielding. After several such pulses the stress under a building, say, will be too small to be able to carry out the loading. If during this loading process the stress state reaches the domain $\partial F/\partial \sigma < 0$, the sand may again become stable, and further loading may significantly increasing both σ and $\bar{\sigma}$.

The stress states in the domain $\partial H/\partial \sigma > 0$ are stable states. However, if the dynamic loading is producing a significant increase of τ (see segment MN in Fig.2), then during this dynamic loading the stress states may still reach states in the strip (7). In this case, repeated pulses during which stress states in the strip (7) or reached, may still produce a loosing of stability, due to the same mechanism.

Experimental evidence

Very nice experimental data obtained by Lade [1993, 1994a,b] have shown that instability of saturated sand in the domain (6) is possible in cyclic loading. The liquefaction was obtained in the laboratory and has taken place in a similar way as described above. This shows that a non-associated viscoplastic model where a compressibility domain exists between the lines $\partial F/\partial \sigma = 0$ and $\partial H/\partial \sigma = 0$ can explain why loosing of stability can take place. In this analysis it is essential to take into account that the viscoplastic yield surfaces are changing very little during a short time loading/unloading pulse. This is impossible in the framework of time-independent plasticity, where the yield surfaces are moving simultaneously with the loading path. For other results concerning saturated sand instability see Zlatovic and Ishihara [1997], Tsukamoto et al. [1998], Uathyakumar [1996], besides others.

Conclusion

A nonassociated elasto/viscoplastic constitutive equation formulated for saturated sand can explain why repeated dynamic loadings can liquefy the sand.

REFERENCES

Anandarajah, A. 1994, Procedure for Elastoplastic Liquefaction Modeling of Sands, *Journal of Engineering Mechanics*, Vol.120, No.7, 1563-1587.

Cazacu, O., J. Jin, and Cristescu, N.D., 1997, A New Constitutive Model for Alumina Powder Compaction. *KONA, Powder and Particle*, no.15, 103-112.

Cristescu, N.D. 1989, **Rock Rheology,** Kluwer Academic Publishers, Dordrecht, 336 pp.

Cristescu, N.D. 1991, Nonassociated elastic/viscoplastic constitutive equations for sand, *Int. J. Plasticity*, Vol.7, No.1, 41-64.

Cristescu, N.D. 1996. Plasticity of porous and particulate materials. Nadai Award Lecture, *Transactions of the ASME, Journal of Engineering Materials and Technology*, Vol.118, 145-156.

Cristescu, N.D., Cazacu, O. and J. Jin , 1997, Constitutive Equation for Compaction of Ceramic Powders. *IUTAM Symposium on Mechanics of Granular and Porous Materials*, Ed. Norman Fleck, Kluwer Academic Publ., pp.117-128.

Cristescu, N.D. and Hunsche, U. 1998. **Time Effects in Rock Mechanics.** *John Wiley and Sons*, Chichester – New York – Weinheim – Brisbane – Singapore – Toronto.

Cristescu, N.D. 1999, Constitutive Equation for Geomaterials. 4th *International Conference on Constitutive Laws for Engineering Materials*. NSF, Rensselaer Polytechnic Institute, Troy, New York, July 27-30, pp.24-27.

Han, C. and Vardoulakis, I.G. 1991, Plane-strain compression experiments on water-Saturated fine-grained sand. *Geotechnique*, Vol. 41, No.1, 49-78.

Ishihara, K. 1993, Liquefaction and flow failure during earthquakes. *Geotechnique*, Vol.43, No.3, 351-415.

Jin, J. and Cristescu, N.D. 1998. A constitutive model for powder materials. Transaction ASME, *Journal of Engineering Materials and Technology*, vol.120, no.2, 97-104.

Lade, P.V., Nelson, R.B., and Ito, Y.M. 1987. Nonassociated Flow and Stability of Granular Materials. *Journal of Engineering Mechanics,* vol. 113, no.9, 1302-1318.

Lade, P.V. 1993. Initiation of static instability in the submarine Nerlerk berm. *Can. Geotech. J., 30*, 895-904.

Lade, P.V. 1994a. Creep Effects on Static and Cyclic Instability of Granular Soils. *Journal Of Geotechnical Engineering,* **120**, 2, 404-419.

Lade, P.V. 1994b. Instability and Liquefaction of Granular materials, *Computers and Geotechnics,* **16**, 123-151.

Nawrocki, P.A., Cristescu, N.D., Dusseault, M.B., Bratli, R.K. 1999. Experimental Methods for Determining the Elastic Moduli for Shale. *Int. J. of Rock Mechanics and Mining Sciences.* Vol. 36, no.5, 659-672.

Nemat-Nasser, S. and Shokooh, A., 1980, A Framework for Prediction of Densification And Liquefaction of Sand in Cyclic Shearing. *Ingenieur – Archiv,* vol.49, 381-392.

Tsukamoto, Y., Ishihara, K. and Nanaka, T. 1998. Undrained Deformation and Strength Characteristics of Soil from Reclaimed Deposits in Kobe. Special Issue of Soils and Foundations, 47-55.

Uathayakumar, M. , 1996, Liquefaction of Sands under Multi-axial Loading. Ph. D. Thesis, University of British Columbia, Vancouver.

Vaid, Y.P. and Sasitharan, S., 1992, The Strength and Dilatancy of Sand. Can. Geotech.J., 29, 522-526.

Veyera, G.E., and Charlie, W.A., 1990, Laboratory Study of Compressional Liquefaction. *Journal of Geotechnical Engineering*, Vol.116, no.5, 790-804.

Zlatovic, S., and Ishihara, K., 1997, Normalized Behaviour of Very Loose Non-plastic Soils: Effects of Fabric. Soils and Foundations, 37 (4), 47-56.

Fundamental Dynamic Behavior of Foundations on Sand

Ronald Y. S. Pak[1], Member, ASCE and Jeramy C. Ashlock[2], Student
Member, ASCE

Abstract

Despite numerous past attempts and research efforts, comparisons of physical test results on the dynamic interaction of foundations with sandy soils with analytical theories have continued to be a subject of considerable debate and inconsistency. To determine the source of the difficulties that besiege the physical problem, a systematic investigation was performed using the centrifuge scaled modeling method to examine at a fundamental level the dynamic behavior of surface foundations under both vertical and horizontal excitations. Finally exposed by the experimental approach, a basic reason for the continuing difficulty in characterizing the dynamic behavior of foundations on sandy soils is discovered to be a key incompatibility of present analytical frameworks with the observed physical soil-foundation behavior. While a deeper physical understanding is required prior to a proper theoretical resolution, a conceptually new but practically simple analytical framework which can explicitly recognize and rationally accommodate the granular soil dynamics problem is found to be possible using the instrument of *Impedance Modification Factors*.

Introduction

The key to assessing the effects of dynamic soil-structure interaction on buildings, bridges and superstructures is a reliable determination of the dynamic response of the foundation under general multi-directional loading. As basic a problem as it can be in soil dynamics, the vibratory characteristics of foundations on granular soils have remained a frustrating mystery for both researchers and practitioners for many years. Physically, a major source of the difficulties probably lies in the complex stress-strain relationship of such deposits under cyclic loading and spatially inhomogeneous stress conditions (e.g., Hardin and Drnevich 1972). Analytically, the solution of the related three-dimensional

---

[1]Professor, University of Colorado, Boulder, CO 80309-0428
[2]Research Assistant, University of Colorado, Boulder, CO 80309-0428

wave propagation problem in an unbounded domain is also a non-trivial mathematical and computational exercise even for elementary material and geometric configurations (e.g. Pak and Gobert 1991). While they may have direct engineering appeal, experimental field testings (e.g. Crouse et al. 1990) are usually limited in their usefulness in deriving definitive fundamental conclusions due to their case-specific nature, uncertainties in site characterization, and their inherent difficulties in achieving significant parametric variations. With the goal towards a rational understanding of the granular soil dynamics problem, a systematic series of vertical vibration experiments on circular footings resting on sandy soils was performed by Pak and Guzina (1995). The result demonstrates that a serious source of past difficulties was an inadequate recognition and incorporation of the influence of foundation bearing pressure, foundation size and soil density on the dynamic response. To provide a direct means for practitioners to deal the problem, they have also synthesized their data base into a formula for an equivalent homogeneous shear modulus for use with the classical half-space solution. To further elevate the current understanding of this class of fundamental soil dynamics problems, a new experimental program was launched to investigate the corresponding case of lateral-rocking vibration. It should be clear however that the lateral-load problem is intrinsically more complicated than the vertical one due to the lack of symmetry in motion, the coupling of the translational and rotational response, and other higher-order contact phenomena (e.g. Hushmand 1983, Novak 1985). In what follows, the experimental procedure used in the investigation will be highlighted, together with some representative results to illustrate the most intriguing aspects of the physical problem. Among other findings, the experimental data exposes a fundamental flaw in current concepts for dealing with the dynamic response of foundations on sandy soils. To cope with such circumstances prior to a complete theoretical resolution, the new concept of *Impedance Modification Factors (I.M.F.)* is introduced whose usefulness to a rational experimental synthesis of this class of dynamic soil-structure interaction problems will be demonstrated.

Experimental Program

1. Soil and Foundation Models

To obtain a realistic simulation and comparison of vertical and lateral dynamic characteristics of foundations resting on granular media, a series of scaled model tests were performed on the 440 g-ton centrifuge at the University of Colorado using a large steel container of rectangular geometry. Lined with Duxeal, an absorptive material, on the inside walls, the container permits the use of a soil model with a plan dimension of 1.0 m by 1.2 m with a height of 0.61 m. As stated in Pak and Guzina (1995), such a configuration was found to be effective in minimizing boundary effects associated with this type of dynamic simula-

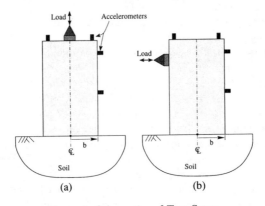

Figure 1: Schematics of Test Setup.

tions. All the footings used in this study had square bases and were made from high strength aluminum. In order to accommodate the necessary transducers such as load cells and accelerometers for both vertical and lateral dynamic tests, holes were drilled and tapped on the top as well as vertical sides of the footing. Schematics of typical model setups are shown in Figure 1. For the purpose of checking the consistency of the scaled modeling approach, model footings were designed to be dimensionally scaled replicas of each other. The dimensions for the smallest footing were 40 mm by 40 mm by 100 mm. Additional footings with the same aspect ratio but different sizes were also fabricated to cover a variety of prototype foundation dimensions. Since the same miniatured transducers were used with all footings and could not be scaled, the instrumented footing's mass (M) and mass polar moment of inertia (Jo) did not scale exactly according to centrifuge scaling relations. However, due to the dominant size and mass of the model footings compared to the transducers, the effects of the constant-size transducers were negligible most of the time according to the results from modeling-of-model tests. As in Pak and Guzina (1995), the soil used in the study was a fine, uniform dry silica sand designated as F-75 by the United States Silica Company. Key physical properties of this sand are listed in Table 1. To create a homogeneous sample, the sand was placed in a hopper and pluviated through air from a pre-determined height from the soil surface. The flux of the sand was controlled by a slotted plate in the bottom of the hopper. For the data presented in this communication, the soil had a relative density of 86%, a mass density of 1731 kg/m^3, and a void ratio of 0.531.

2. Instrumentations and Setup

To generate vertical and horizontal forced excitations, 2 Bruel and Kjaer

Table 1: Physical Properties of F-75 Silica Sand

Property	Description
Mineral	Quartz
Grain Shape	Rounded
Specific Gravity	2.65
Surface Area $(cm^2/gr))$	162

(B & K) electromagetic exciters 4809 and 4810 were employed respectively. They both have a frequency range of 10-20000 Hz which was more than adequate for the spectrum of interest. To have a precise point of application and measurement of the dynamic force transmitted, a stinger design which could be attached to the model footing was developed to incorporate a miniaturized washer-shaped Kistler 9001 load cell. To monitor the motion of the foundation, PCB model 303A11 and PCB352B67 miniaturized accelerometers with ultra-low transverse-sensitivity were used. As in Pak and Guzina (1995), the method of random excitation was employed in the dynamic testing because of its efficiency and reliability. With the aid of a Tektronix 2630 Fourier analyzer which could generate random excitations, process the analog response time-history signals with averaging and windowing, and determine the frequency response functions in a forced vibration problem, accelerance functions which are ratios of accelerations at selected locations to the applied dynamic load in the frequency domain were used for the dynamic characterization of the soil-structure interaction problem.

Experimental Results and Synthesis

1. Vertical Response

To ascertain the validity of the observations of Pak and Guzina (1995) for square foundations, the aforementioned scaled model footings with square bases were tested under random vertical excitation from the B & K 4809 exciter at different g-levels to simulate different prototype foundation dimension and contact stress levels. In all cases, the dynamic characteristics of the square foundation were found to be very similar to those found in Pak and Guzina (1995) for a circular footing. By matching a rigorous boundary element solution (Pak and Guzina 1999) with a Poisson's ratio $\nu = 0.25$ for a square foundation on a homogeneous half-space with the experimental data, an equivalent homogeneous shear modulus $G_{eq.hom.}$ for each model test with its specific average static contact pressure p_{pr} and foundation half-width b_{pr} was determined. As an illustration, a typical accelerance function at the top of a square foundation under central vertical excitation is shown in Fig. 2 together with the calibrated continuum solution. As can be seen from the figure, the quality of agreement is comparable to those in Pak and Guzina (1995), confirming the usefulness of the homogeneous half-space model for the vertical vibration problem. Furthermore, by synthesizing the equivalent homogeneous shear modulus as a function of the average soil bearing pressure and foundation dimension, it is found that the $G_{eq.hom.}$s for all the tested square foundations can be well summarized by the power law format

$$G_{eq.hom.} \; = \; G' \; (\frac{b_{pr}}{1m})^{0.1}(\frac{p_{pr}}{1MPa})^{0.5}, \tag{1}$$

where the constant G' has a value of 6.31 MPa for the particular soil density. By virtue of Eq. (1), the equivalent homogeneous shear modulus can be determined for any square foundation configuration including those used in the lateral vibration experiments. Besides its confirmed usefulness in modeling vertical vibration, the calibrated homogeneous half-space model is of critical importance in exposing the characteristics of the lateral mode of foundation vibration and the inadequacy of current concepts for a general treatment of this class of problems.

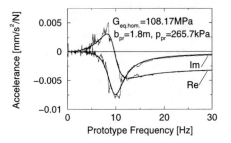

Figure 2: Vertical test: experimental vs theoretical accelerance from an equivalent homogenous half-space

2. Lateral Response

To facilitate a direct comparison with the vertical vibration response, the model footing configurations used in the lateral tests were basically identical to those in the vertical tests. The only difference was in the slight reduction in the bearing pressure due to the slightly different accelerometer arrangements and the absence of the armature weight coming from the B & K 4809 exciter for vertical vibration tests. In all the lateral tests, the random forcing was supplied by the B & K 4810 exciter which was mounted horizontally next to the footing. Shown in Fig. 3 is a typical experimental accelerance function generated from a lateral vibration test for a side location on the same footing used in the vertical test in Fig. 2 under comparable conditions. One can see that the curve has a sharp peak at low frequency and is followed by more gradual variations as the frequency increases. To illustrate the issues at hand, also included in the figure is the theoretical accelerance generated by the horizontal and rocking impedances K_{hh} and K_{mm} from the continuum solution of a rigid square foundation on a homogeneous half-space with the shear modulus given by Eq. (1). A careful examination of Fig. 3 would reveal that while the overall shape and trend of the theoretical and experimental accelerance functions are comparable, both the sharp peak and the gradual portion of the experimental accelerance curves are noticeably off from the theoretical curves. As was often done in past studies (e.g. Hushmand 1983, Brown 1995) in the absence of vertical vibration data, one could ignore the vertical result and determine a separate shear modulus $G_{eq.hom}$ so that the homogeneous half-space solution or Lysmer's analog will match with the experimental curve from the lateral test. Such an approach is shown in Fig. 4. While the agreement is now significantly better between the theoretical solution and the measured data, the necessary shear modulus in the re-calibrated theoretical solution is found to be 76.9 MPa, a value which is significantly lower than the modulus of 105.3 MPa from vertical vibration. The foregoing observation has also been found in earlier experimental results in Brown (1995) for circular footings subjected to lateral dynamic loads. In that study, it was noted that the equivalent homogeneous shear modulus for the case of a circular surface foundation loaded laterally ranged from being 27 to 37 percent smaller than that predicted by Pak and Guzina's equation. While using different shear moduli to characterize the same foundation soil for different mode of vibration may be considered as viable from the viewpoint of expediency, its clear conflict with common sense and intuition reflects the inherent complexity of the subject and demonstrates the need of a better understanding and accomodation of the physical problem.

Concept of Impedance Modification Factors

To synthesize the experimental results without the use of a nonunique mechanical characterization of the soil, a new conceptual understanding and analytical framework which can accomodate the observed characteristics without violating common sense and engineering intuition is clearly needed. To this end, it is important to first accept the possibility that the classical homogeneous half-space footing solution from which Lysmer's simplified mechanical analog was developed may not fully describe the key characteristics of a footing founded on a cohesionless soil under dynamic loading. One well-known deficiency of the classical solution is its prediction that the contact stress between a rigid footing and the soil will reach a local minimum at the center, when in reality a granular subgrade will produce a local maximum at the said location. Due

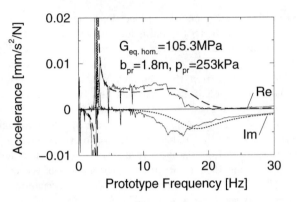

Figure 3: Lateral test: expermental vs theoretical acceleration with $G_{eq.hom.}$ from vertical test

to the soil's self weight, the shear modulus of a uniform sand is also known to vary with depth in reality instead of being a constant throughout the medium. Furthermore, immediately under the footing, the increase in confining pressure will also lead to a locally stiffened soil modulus which is not accounted for in the continuum solution. Because of the neglects of these complicated but characteristic aspects, it should therefore not be totally surprising in hindsight that the classical theoretical solution cannot be made to deliver good agreements for all modes of vibration simultaneously. To cope with the deficiencies of the current standard approach without relying on advanced theoretical solutions, a new but simple analytical framework that is found to be effective for synthesizing the physical problem rationally is to revise the foundation impedance matrix representation to a new form of

$$\left[\, K_{ij} \, \right]_{(sand)} = \left[\begin{array}{ccc} K_{vv}(\omega) & 0 & 0 \\ 0 & \alpha_{hh}K_{hh}(\omega) & K_{hm}(\omega) \\ 0 & K_{mh}(\omega) & \alpha_{mm}K_{mm}(\omega) \end{array} \right]. \quad (2)$$

Here, K_{ij} are theoretical impedances from the classical homogeneous half-space solution whose shear modulus is determined by matching K_{vv} with the experimentally determined vertical impedance of the foundation. As discussed in Pak and Ashlock (2000), the coefficients α_{ij} in (2) are called *Impedance Modification Factors (I.M.F.)* for obvious reason. Created to represent the part of the physical characteristics of the soil-foundation system not captured in the chosen theoretical anchor, the *I.M.F.*s can be derived from experimental measurements. Re-examining the experimental data in Fig. 3 in the new conceptual and analytical framework of Eq. (2) for instance, one can determine the optimal pair of

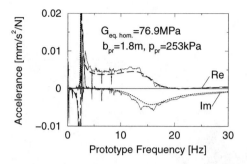

Figure 4: Lateral test: experimental vs theoretical accelerance with an alternative $G_{eq.hom.}$

α_{hh} and α_{mm} that will minimize the least-square error of the functional match between the experimental data and the new theoretical accelerances. The result of such an analysis is shown in Fig. 5 which demonstrates the existence of the optimal α_{hh} and α_{mm} for the illustrative example. Noteworthy here is that (i) α_{hh} is generally not equal to α_{mm} and (ii) they are both less than 1, indicating that the rocking and horizontal impedances are smaller than those predicted by an equivalent homogeneous half-space model. The full performance of the impedance-modification-factor approach is illustrated in Fig. 6 where the results in Fig. 4 are plotted against the new theoretical solution in the frequency spectrum. As one can see from the display, the quality of the match generated by the Impedance-Modification-Factor approach is a further improvement over the alternate-shear-moduli method. Most importantly, the proposed analytical approach constitutes a new conceptual framework which is apt to be the prerequisite to a rational resolution for this class of soil dynamics problems.

Summary

In this paper, the result of a fundamental experimental investigation of the problem of dynamic foundation-soil interaction on granular soils is highlighted. Through a direct comparison of vertical and lateral vibrations of foundations on sand with each other and contrast them with current analytical methods, the salient features and the underlying challenge of the soil dynamics problem are exposed vividly. To deal with the serious incompatibility of the current theoretical approach with the physical problem, a conceptually new but practically simple analytical framework which can explicitly recognize and rationally accommodate the granular soil dynamics problem is found to be possible by virtue of the method of *Impedance Modification Factors*.

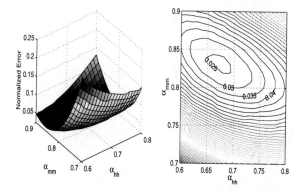

Figure 5: Dependence of Best-Match Least-Square Error on Impedance Modification Factors

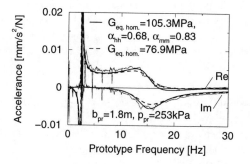

Figure 6: Lateral test: expermental vs theoretical accelerance with α_{hh}, α_{mm} and $G_{eq.hom}$

Acknowledgements

The support from the National Science Foundation through Grant CMS 9320539 and CMS 9712835 is gratefully acknowledged.

References

Brown, E.C. (1995). "Centrifuge modeling of surface foundations subject to dynamic loads," M.S. thesis, University of Colorado, Boulder.

Crouse, C.B., Hushmand B., Luco, L.E., and Wong, H.L. (1990). "Foundation impedance functions: theory versus experiment," J. Geotech. Engrg., ASCE, 116(3), 432-449.

Hardin, B.O. and Drnevich, V.P. (1972). "Shear modulus and damping in soils: design equations and curves," J. Soil Mech. Found. Engrg., ASCE, 98, 668-692.

Hushmand, B. (1983). "Experimental studies of dynamic response of founda-tions," doctoral thesis, California Institue of Technology, Pasadena.

Novak, M. (1985). "Experiments with shallow and deep foundations," Vibration Problems in Geotechnical Engineering, Gazetas and Selig (ed.), ASCE, 1-26

Pak, R.Y.S. and Gobert, A.T. (1991). "Forced vertical vibration of rigid discs with arbitrary embedment," J. Engrg. Mech., ASCE, 117 (11), 2527-2548

Pak, R.Y.S. and Guzina, B.B. (1995). "Dynamic characterization of vertically loaded foundations on granular soils," J. Geotech. Engrg., ASCE, 121 (3), 274-286.

Pak, R.Y.S. and Guzina, B.B. (1999). "Seismic soil-structure interaction anal-ysis by direct boundary element methods," Intl. J. Solid. Struct., 36 (31-32), 4743-4766.

Pak, R.Y.S. and Ashlock, J.C. (2000). "Concept of Impedance Modification Factors for foundations on cohesionless soils," Proc. 14th Engrg. Mech. Conf., Austin, Texas.

Comparison of Deep Foundation Performance in Improved and Non-improved Ground using Blast-Induced Liquefaction

Scott A. Ashford[1], Member, Kyle M. Rollins[2] Member, and Juan I. Baez[3], Member

Abstract

The results presented in this paper were developed as part of a larger project analyzing the behavior of full-scale laterally loaded piles in liquefied soil, the first full-scale testing of its kind. This paper presents the results of a series of full-scale tests performed on deep foundations in liquefiable sand, both before and after ground improvement, where controlled blasting was used to liquefy the soil surrounding the foundations. Data were collected showing the behavior of laterally loaded piles before and after liquefaction. After the installation of stone columns, the tests were repeated. Based on the results of these tests, it can be concluded that the installation of stone columns can significantly increase the density of the improved ground as indicated by the cone penetration test. The stone columns were found to significantly increase the stiffness of the foundation system, by more than 2.5 to 3.5 times that in the liquefied soil. However, in non-liquefied ground, the improvement from stone columns could be more than compensated for by increasing the piles. In liquefied soil, however, more than doubling the number of piles or increasing shafts diameters by 50 percent did not nearly match the improved performance of the treated ground. This study provides some of the first full-scale quantitative results on the improvement of foundation performance due to stone columns in a liquefiable deposit.

Introduction

Installation of stone columns, also referred to as vibro-replacement and vibro-displacement, is a ground improvement technique often used to mitigate liquefaction hazards in saturated loose granular soils. Stone columns can improve the

---

[1] Assistant Professor of Geotechnical Engineering, Department of Structural Engineering, University of California, San Diego, 9500 Gilman Drive, La Jolla, CA 92093-0085

[2] Professor of Geotechnical Engineering, Department of Civil and Environmental Engineering, Brigham Young University, 386S, Provo, UT 84602

[3] Chief Engineer and Regional Operations Manager, Hayward Baker Inc., 1780 Lemonwood Drive, Santa Paula, CA 93060

performance of these deposits in four main ways. First, the installation of the stone columns densifies the deposit by vibration and replacement. Second, this technique increases the lateral stresses in the surrounding soil. Third, stone columns provide reinforcement, as the stone columns are stiffer, stronger and denser than the surrounding soils. Finally, stone columns provide drainage, reducing the potential for build-up of excess porewater pressures. Though the effect of each of these factors will vary between deposits, combined they make stone columns an efficient and popular liquefaction hazard mitigation technique (Kramer, 1996).

There is considerable qualitative data available showing that stone column installation is an effective means of ground improvement for mitigating liquefaction hazards. A report by Mitchell et al. (1995) gives case histories for the performance of improved ground during earthquakes for more than 30 sites. Five of these sites, from the 1989 Loma Prieta Earthquake and the 1994 Northridge Earthquake, were treated with stone columns. In each case, good performance was observed following the earthquake. Priebe (1990) describes additional case histories for improved sites that performed well in earthquakes. Other reports have discussed the design of stone column ground improvement and the extent of improvement that can be expected from the use of stone columns (e.g. Somasumdaram et al., 1997; Saxena and Husin, 1997; Soydemir et al., 1998).

While qualitative information from past earthquakes is valuable in confirming that stone columns can be an effective means of ground improvement, quantitative data is needed for design purposes. This paper is a result of a series of full-scale tests performed on deep foundations in liquefiable sand, before and after ground improvement, and were part of a larger project analyzing the full-scale behavior of laterally loaded piles in liquefied soil, the first full-scale testing of its kind. Controlled blasting was used to liquefy the soil surrounding the piles, and data was collected showing the behavior of laterally loaded piles before and after liquefaction. After the installation of stone columns, the tests were repeated. This paper presents a comparison between the performance deep foundations in ground improved by stone columns to that of the same and larger (i.e. larger diameter or more piles) foundation systems in unimproved ground, both before and during liquefaction.

Project Description

The stone column tests presented in this paper were part of a larger series of tests on the full-scale behavior of laterally loaded piles in liquefied sand. This project, known as the Treasure Island Liquefaction Test (TILT), was a joint venture between the University of California, San Diego and Brigham Young University. As part of the TILT project, a 4- and 9-pile group of 324-mm diameter pipe piles was loaded laterally against a 0.6-m and 0.9-m Cast-In-Steel-Shell (CISS) pile, respectively in 15- by 21-m excavations, approximately 1.5-m deep. A high-speed hydraulic actuator was used to apply the lateral loads. The first lateral load test of the 4-pile group and 0.6-m CISS pile was conducted before the installation of stone columns and before blasting. Liquefaction was then induced using controlled blasting, and the piles were tested again. Stone columns were then installed, and the tests were repeated for both pre-blast and post-blast behavior. Load-displacement and pore pressure information was gathered for all tests. Similar testing was carried out at an adjacent excavation on the 9-pile group and 0.9-m CISS pile without the installation of stone columns.

The testing took place on Treasure Island, a man-made island built on a shoal of Yerba Buena Island in the San Francisco Bay. Treasure Island has been a naval base for a number of years, but is currently being transferred from the U.S. Navy to the City of San Francisco as part of a national base closure program. The site at Treasure Island was selected for a number of reasons. Approvals for the use of explosives were relatively easy to obtain for the portion of the island still operated by the U.S. Navy. In addition, the site is only 300 meters away from the Treasure Island Fire Station, the location of a National Geotechnical Experimentation Site (NGES). The NGES status of the site, as well as numerous other geotechnical investigations on the island, provides a wealth of geotechnical data to draw from for the TILT project. Furthermore, there is a known liquefaction hazard at the site due to the high groundwater level and loose nature of the hydraulic fill. In fact, liquefaction was observed across the island during the 1989 Loma Prieta earthquake (Andrus et al., 1998).

The generalized soil profile at the project site is shown in Figure 1, along with values for the Standard Penetration Test (SPT) $N_1(60)$ values. The groundwater level was at a depth of approximately 1.5 meters. The soil profile generally consists of uniformly graded sand with silt to a depth of approximately 6 meters. Sieve analyses shows the fines content ranges between 5 and 15 percent. This would classify the soil as SP-SM based on ASTM D-2488. This surficial sand is the layer where all of the controlled blasting took place. It is underlain by soft fat clay (CH) to a depth of approximately 10 meters, loose silty sand (SM) to 13.5 meters in depth, and again soft fat clay to the bottom of the borehole at 19 meters. Though the top 1.5 meters of soil at the site was excavated for the experiments, all depths given are from the original ground surface unless indicated otherwise.

Methods and Testing Procedure

The field portion of this project lasted between September 1998 and June 1999. All of the lateral load tests were carried out in February 1999. Below is a description of the test set-up, controlled blasting and testing procedures, as well as a description of the stone column installation.

Test Set-up

Figure 2 presents a plan view of the test set-up. The 4-and 9-pile groups consisted of 342-mm O.D. steel pipe piles with a 10-mm wall thickness, connected by a load frame that allowed for the free rotation of the pile heads while maintaining the same lateral displacement for all four piles in the group. The CISS piles were 0.6 m and 0.9 m in diameter, with nominal wall thicknesses of 13 mm and 11 mm, respectively. The hydraulic actuator used had a double swivel connection to both the CISS pile and pile group load frame thus allowing for free rotation at the CISS load stub. All of the piles were driven to depths between 12 and 14 meters below the excavated surface. For the CISS pile, the steel shell was driven into place, drilled out, and then filled with steel reinforcement and concrete. No water was observed in the steel shell prior to placement of the concrete.

The sites were excavated to a depth of approximately 1.5 meters. The objective of this was to conduct the lateral load test primarily in the loose saturated sand by removing the medium dense sand and lowering the excavated surface closer to the ground water table. Prior to excavation, SPTs and CPTs were performed. Following completion of the pre-treatment load tests, the 4-pile group/0.6-m CISS pile site was backfilled to the original elevation, and the stone columns were installed. A second set of CPTs was then performed. The site was then re-excavated to the same level as before, and the post-treatment series of tests were performed.

A 1500-kN high-speed hydraulic actuator was used to laterally load the piles, with the loading point approximately 0.8 meters above the excavated surface. The speed of the actuator was approximately 10 mm/second. For each case, the actuator was connected between the load frame of the pile group and the load stub of the CISS pile, such that load-displacement information for the pile group and the CISS pile was obtained simultaneously. The applied load was measured in the actuator using an array of three 500-kN load cells. Relative displacement between the pile group and CISS pile was measured in the actuator as well, using a linear variable displacement transducer (LVDT). In addition, absolute displacement measurements of the piles were obtained using string activated linear potentiometers fixed to a reference post outside of the excavation. These were attached to the foundations in such a way as to allow for monitoring of displacement, tilt, and rotation. In order to monitor the effectiveness of the controlled blasting, pore pressures were measured using piezoceramic pore pressure transducers arranged at various depths immediately adjacent to the piles.

Controlled Blasting

In order to determine the blasting procedure to be used at the site, a pilot liquefaction study was performed on an adjacent site. This pilot study confirmed that liquefaction could be induced by controlled blasting techniques. For safety reasons, two-part explosives were used on the project. When mixed, the nitromethane and ammonium nitrate had the equivalent explosive power of 0.5 kg of TNT (trinitrotoluene) per charge. The explosives used to raise pore pressures in the vicinity of the piles consisted of two sets of eight 0.5-kg charges, placed in 5-m diameter circles around each of the piles. All charges were placed at a depth of approximately 3.5 meters below the excavated ground surface. The sixteen charges were set off in one sequence, with a delay of approximately 250 milliseconds between each pair of charges. For the post-treatment blast, the same charge pattern was used. In addition, 14 charges were added around the perimeter of the improved ground excavation in an attempt to induce liquefaction in the area surrounding the stone columns. These additional charges had no effect on the soil immediately surrounding the foundations, but this was an attempt to study the behavior of an "improved island" and will be discussed in a future publication.

Stone Column Installation

After the first series of tests, the site was backfilled and stone columns were installed around the piles. Twenty-four 0.9-m diameter stone columns were installed in a 4 by 6 grid around the piles, with a spacing of approximately 2.4 meters on center, as shown in Figure 2. The stone columns extended through the surficial sand layer at the site, to depths of approximately 5.5 to 6.0 meters. After the installation, the site was re-excavated to the same depth as for the pre-treatment tests.

The stone columns used were installed using the dry bottom feed method. "Dry" in this context refers to the fact that the vibratory probe was driven into the ground using compressed air instead of water. The term "bottom feed" is in reference to the way gravel is fed through the tip of the probe rather than being placed into the soil from the ground surface. Compressed air, vibration, and the weight of the probe itself drove the probe into the ground. Once the probe reached full depth, it was lifted up and the hole was backfilled with gravel from the probe tip. The probe was approximately 0.5 m in diameter, and required multiple passes to create a column 0.9 m in diameter. The probe was raised in 0.9-m lifts, and gravel was placed into the soil. The probe was then re-lowered into the gravel that had just been placed, forcing it outward and further densifying the surrounding soil. Lifting the probe multiple times inserts more gravel into the column. To determine the number of passes required for complete site treatment, the operator monitored the amperage of the vibrating probe. As the soil was densified, the probe required more power to maintain its vibration. Once a set level of increase in amperage had been reached, the operator proceeded to the next 0.9-m lift.

Loading Sequence

All foundations were loaded prior to blasting in order to obtain baseline information in the non-liquefied state. In the case of the 4-pile group/0.6-m CISS pile test, a complete series of tests were conducted prior to installation of stone columns. For the pre-liquefaction tests, the piles were pulled towards each other until one pile was displaced 38 mm. The load was reduced until one of the piles returned to its original position. After this test, the charges were set off. Ten seconds after the blast, the piles were loaded again, cycled under displacement control to 75 mm, 150 mm, and 225 mm of absolute displacement, then cycled at 225 mm of displacement nine times. For these tests, the load level was approximately 0.8 meters above the excavated surface.

The procedure for the post-treatment tests was essentially the same as for the pre-treatment testing. After the first tests were completed, stone columns were installed and the post treatment testing took place. For the post-treatment testing, the same loading sequence as the pre-treatment tests was attempted. However, the capacity of individual load cells within the pile group was exceeded before the piles had reached 150 mm of absolute displacement, so the piles were cycled under load control instead of displacement control.

Results

The improvement to the upper sand layer is apparent from review of Figure 3, which shows the CPT tip resistance values (q_c) for the upper sand layer, both before (Figure 3a) and after (Figure 3b) treatment with stone columns. Excluding the top 1.5 meters that was excavated prior to testing, a substantial increase in the tip resistance can be seen throughout the sand layer. Prior to installation of the stone columns, the average tip resistance in the upper sand was approximately 4 MPa. After installation of the stone columns, the average tip resistance in the upper sand ranged between 10 and 50 MPa, and below a depth of 2 meters (i.e. 0.5 meters below the excavated surface) the average is well above 20 MPa. This amount of improvement can be expected from the installation of stone columns (e.g. Priebe, 1991; Soydemir, 1997). Clearly, this substantial increase in tip resistance corresponds to a substantial decrease in the susceptibility of the upper sand to liquefaction.

This increased resistance to liquefaction was observed in comparison of the pre- and post-treatment excess pore pressures (Ashford *et al.*, 2000). Though space limitations prevent presentation of the detailed observations, a summary of their findings is presented here. Ashford *et al.* (2000) found an immediate increase in pore pressure at all depths at the time of the blast and these were maintained generally in excess of R_u equal to 80 percent for depths greater than 2.7 meters for the first 10 minutes of loading. Though R_u was not found to be 100 percent throughout the profile, observations confirmed that the site was essentially liquefied. These observations included numerous sand boils, water flowing freely from the ground, and considerable surface settlement (Ashford and Rollins, 1999). Ashford *et al.* (2000) found the pore pressure response for the post-treatment tests in sharp contrast to those recorded before installation of the stone columns. Though a sudden increase in pore pressure was apparent at the beginning of the record following the blast, the increase was much less than in the pre-treatment case ($R_u = 60\%$). Furthermore, rapid dissipation of excess pore pressures was observed. For example, at the end of 10 minutes, pore pressures were nearly hydrostatic, and in fact were slightly negative in some cases. Observations were consistent with these measurements, in that no visible signs of liquefaction were apparent. No sand boils, surface settlement, or flowing water were observed, and there was actually significant gapping around the piles during the cyclic loading.

Perhaps the most dramatic indicator of soil improvement as a result of stone column installation, and of direct importance to this study, are the load-displacement curves. Reviewing first the pre-treatment plots for both the 4-pile group and the 0.6-m CISS pile, shown respectively in Figures 4 and 5, the pre-blast secant stiffness from the plots is approximately 7.5 kN/mm. These values are immediately reduced by over 60 percent due to the increased pore pressure from the blast. As the number of cycles increase and the soil structure is broken down, the secant stiffness is further reduced a total of nearly 70 percent for the 4-pile group and 80 percent for the CISS pile. Higher excess pore pressures observed surrounding the CISS pile may explain the slightly lower stiffness values.

Similar results are observed for the non-treated soil for the 9-pile group and 0.9-m CISS pile shown in Figures 6 and 7, respectively. The pre-blast secant stiffness is approximately 14.3 kN/mm for the 9-pile group and 20 kN/mm for the 0.9-m CISS pile. After several cycles of post-blast loading, both of these values reduce to approximately 3.5 kN/mm, a decrease of over 75 to 80 percent from the pre-blast case.

In sharp contrast are the post-treatment load-displacement curves for the 4-pile group and the 0.6-m CISS pile, shown in Figures 8 and 9, respectively. The initial secant stiffness prior to blasting is approximately 9.3 kN/mm for the 4-pile group and 10.8 kN/mm for the CISS pile. This is an increase of 25 to 45 percent over the pre-treatment, pre-blast test results. For both foundation types, however, the post-blast secant stiffness is approximately 7 kN/mm. This is only a 25 to 35 percent decrease from the pre-blast values.

It is of interest to compare the test results in the improved ground to both cases of non-improved ground. In a direct comparison of the pre- and post-treatment tests for the 4-pile group, it can be seen that the pre-blast secant stiffness is increased by 25 percent as a result of ground improvement. Similarly, a 45 percent increase is observed for the 0.6-m CISS pile. A much more dramatic improvement is observed post-blast, where the improved ground yields a secant stiffness 2.5 times greater for the 4-pile group and 3.5 times greater for the 0.6-m CISS pile.

A comparison of the test results of the post-treatment 4-pile group and 0.6-m CISS pile to those for the non-treated 9-pile group and 0.9-m CISS pile gives an indication of the effectiveness of increasing the pile size or number of piles in lieu of ground treatment. This comparison shows a more substantial foundation may be worthwhile in the non-liquefied case. The secant stiffness for the non-treated 9-pile group is over 50 percent higher than the treated 4-pile group. The increase is over 80 percent when comparing the non-treated 0.9-m CISS pile to the treated 0.6-m CISS pile. However, the post-blast comparison is more favorable to the improved ground case. When comparing the treated 4-pile group to the non-treated 9-pile group and the treated 0.6-m CISS pile to the 0.9-m CISS pile, the treated ground in both cases yields a secant stiffness twice that of the unimproved case.

It is understood that many factors influence the comparison between a more substantial foundation and ground improvement. These factors are not limited to the details of the foundation system, the soil profile, and construction considerations. However, in this study of full-scale test results in liquefied ground, it was shown that more than doubling the number of piles (from 4 to 9) or increasing the shaft diameter by 50 percent (from 0.6 to 0.9 m) does very little for foundation performance during liquefaction in comparison to ground improvement by stone columns.

Conclusions

This paper presents the results of full-scale lateral load tests in liquefiable sands before and after ground improvement with stone columns. In each case, controlled blasting was used to elevate pore pressures in an attempt to liquefy the soil surrounding the deep foundations. Based on these results, several conclusions can be made regarding the stone column effectiveness in improving the performance of the foundation system under lateral loading in liquefiable soils.

- As has been observed in previous studies, the installation of stone columns significantly increased the density of the ground surrounding the foundations as indicated by the cone penetrometer test.
- The installation of stone columns significantly increased the stiffness of an identical foundation system before and after blasting. This increase was 2.5 to 3.5 times that of the system in the liquefied soil.
- Increasing the number of piles in a group from 4 to 9 or increasing the diameter of CISS piles from 0.6 to 0.9 m more than compensated for the ground improvement in the non-liquefied case.
- Increasing the number of piles in a group from 4 to 9 or increasing the diameter of CISS piles from 0.6 to 0.9 m still resulted in much lower foundation stiffnesses as compared to the site improved with stone columns for the post-blast (liquefied) case.

Acknowledgments

The authors wish to express their sincere gratitude to the sponsors of the TILT project: Caltrans (lead agency), Alaska DOTPF, Missouri DOT, Oregon DOT, Utah DOT, and Washington State DOT. In addition, several companies donated services to the project including Hayward Baker Inc., Geotechnics America/Mustang Construction, Condon–Johnson & Associates, Foundation Constructors, Subsurface Consultants, and Kleinfelder & Associates. Finally, the cooperation of the U.S. Navy and the City of San Francisco is greatly appreciated.

References

Andrus, Ronald D., Kenneth H. Stokoe, II, Riley M. Chung, and James A. Bay. *Liquefaction Evaluation of Densified Sand at Approach to Pier 1 on Treasure Island, California, Using SASW Method.* Report NISTIR 6230. National Institute of Standards and Technology, 1998.

Ashford, S.A., and Rollins, K.M., *Full-Sclae Behavior of Laterally Loaded Deep Foundations in Liquefied Sand: Preliminary Test Results,* Structural Systems Research Project Report No. TR-90/03, June 1999, 78 p.

Ashford, S.A., Rollins, K.M., Bradford, S.C., Weaver, T.J., and Baez, J.I., "Liquefaction Mitigation using Stopne Columns around Deep Foundations: Full Scale Test Results," *TRB 79th Annual Meeting CD-ROM,* Transportation Research Board, 2000, 22 p.

Kramer, Steven L. *Geotechnical Earthquake Engineering.* Prentice-Hall, Inc., New Jersey, 1996.

Mitchell, James K., Christopher D. P. Baxter and Travis C. Munson. Performance of Improved Ground During Earthquakes. *Soil Improvement for Earthquake Hazard Mitigation*, American Society of Civil Engineers, Geotechnical Special Publication No. 49, 1995, pp. 1-36.

Priebe, H. J. The prevention of liquifaction by vibro replacement. *Proceedings of the International Conference on Earthquake Resistant Construction and Design.* S. A. Savidis, editor. A. A. Balkema Publishers, Rotterdam, Netherlands, 1991, pp. 211-219.

Saxena, D. S. and J. D. Hussin. Stone Column Improved and Piezocone Tested Site Supports Mid Rise Building Complex – A Case History. *Ground Improvement, Ground Reinforcement, Ground Treatment, Developments 1987 – 1997*, American Society of Civil Engineers, Geotechnical Special Publication No. 69, 1997, pp. 476-491.

Somasundaram, Suji, Gamini Weeratunga, and John Best. Ground Improvement at the Long Beach Aquarium of the Pacific – A Case Study. *Ground Improvement, Ground Reinforcement, Ground Treatment, Developments 1987 – 1997*, American Society of Civil Engineers, Geotechnical Special Publication No. 69, 1997, pp. 457-475.

Soydemir, Cetin, Frank J. Swekosky, Juan I. Baez and Joel S. Mooney. Ground Improvement at Albany Airport, New York. *Ground Improvement, Ground Reinforcement, Ground Treatment, Developments 1987 – 1997*, American Society of Civil Engineers, Geotechnical Special Publication No. 69, 1997.

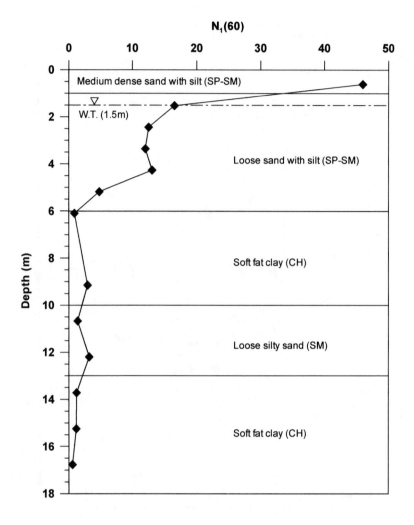

Figure 1: Soil profile and standard penetration test results.

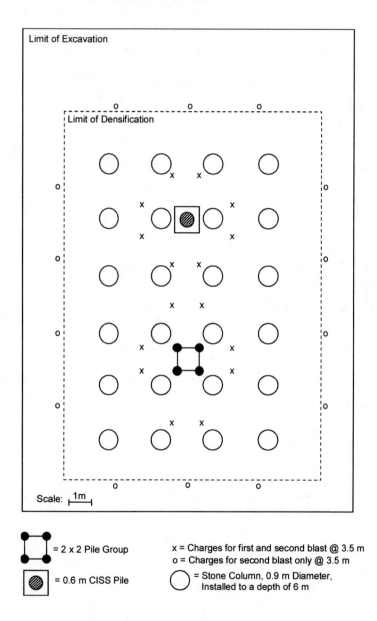

Figure 2: Plan View of test set-up.

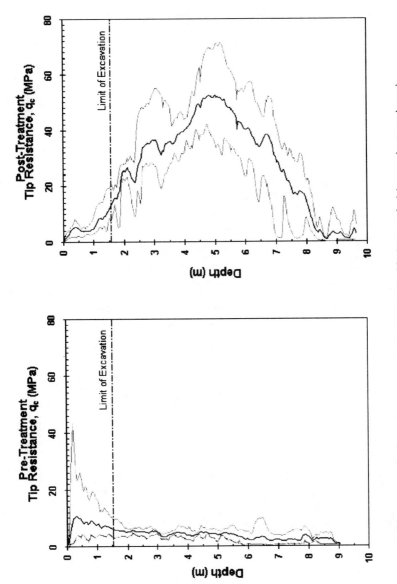

Figure 3: Pre-and post-treatment cone penetration test results. Maximum and minimum values are shown in grey, average values in black.

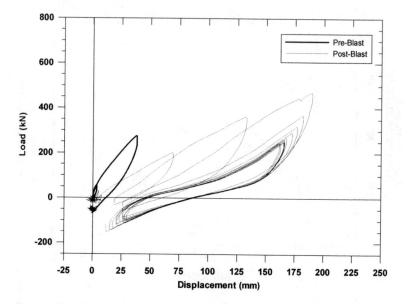

Figure 4: Load-displacement curve for 4-pile group (pre-treatment).

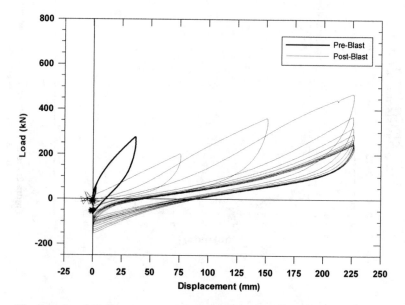

Figure 5: Load-displacement curve for 0.6-m CISS pile (pre-treatment).

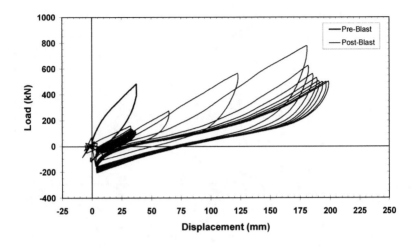

Figure 6: Load-displacement curve for 9-pile group (non-treated).

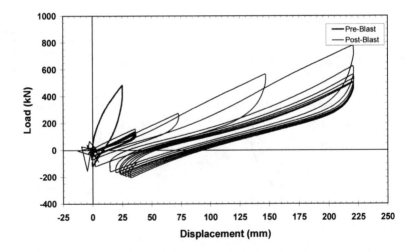

Figure 7: Load-displacement curve for 0.9-m CISS pile (non-treated).

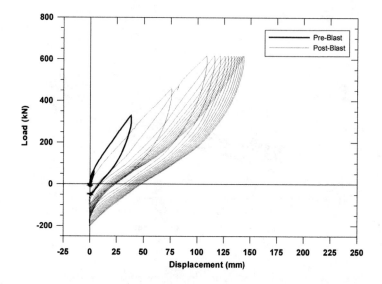

Figure 8: Load-displacement curve for 4-pile group (post-treatment).

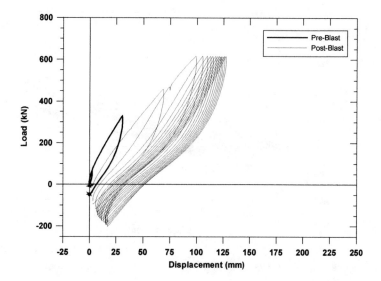

Figure 9: Load-displacement curve for 0.6-m CISS pile (post-treatment).

The influence of high confining stress on the cyclic behavior of saturated sand

R Scott Steedman[1], Richard H Ledbetter[2] and Mary Ellen Hynes[2]

Abstract

This paper describes the findings of an experimental study supported by the U.S. Army Centrifuge Research Center and Engineer Earthquake Engineering Research Program (EQEN) into the behavior of saturated sands under high initial effective confining stresses subjected to strong ground shaking. The research was conducted using the Army Centrifuge at the U.S. Army Engineering Research and Development Center (ERDC), located in Vicksburg MS, formerly known as the Waterways Experiment Station (WES). The centrifuge studies have shown that the generation of excess pore pressure is capped at a level below 100 percent for vertical effective confining stresses exceeding around 3 atmospheres (atm, or 300 KPa). This limit reduces at higher confining stresses. If verified, the potential benefits from this finding for the design of remediation works for large earth dams could be substantial. The paper describes the equipment used for the experiments, the research program, and presents the initial results, contrasting the development of excess pore pressure at low confining stress with that at high confining stress. Possible consequences for a hypothetical dam are discussed.

1.0   Introduction

The current state-of-practice for the evaluation of liquefaction potential and for remediation design and analysis depends on empirical correlations of in-situ measurements of strength versus field experience of liquefaction at shallow depth and laboratory data of the behavior of confined elements under cyclic loading. (Liquefaction is defined here to mean the development of pore pressure equal to 100% of the initial vertical effective stress.) This approach is known as the "simplified procedure". Opinions vary as to the maximum depth in the field at which liquefaction

---

[1] GIBB Ltd, LAWGIBB Group, Reading, UK
[2] US Army ERDC Waterways Experiment Station, Vicksburg, MS

has been observed, but there is no established field evidence from historic earthquakes of liquefaction at depths greater than a few tens of meters. The NCEER Workshop in 1996 on the Evaluation of Liquefaction Resistance of Soils noted that the simplified procedure was developed from evaluations of field observations and field and laboratory test data, Youd and Idriss (1997). The report notes, "These data were collected mostly from sites ... at shallow depths (less than 15m). The original procedure was verified for and is applicable only to these site conditions".

Hence, in design practice the assessment of liquefaction under high initial effective confining stress, such as might relate to the foundations of large earth dams, is based on the extrapolation of observed behavior and correlations at shallow depths. In practice, the behavior of saturated soil under these conditions is not well understood. Based on the results of laboratory tests, researchers have postulated that there is a reduction in the liquefaction resistance of such soil compared to shallow depths. This reduction is accounted for in standard approaches by a ratio known as K_σ, a "correction factor" developed by Seed (1983) in the simplified procedure. This strength ratio is postulated to reduce with increasing initial effective confining stress, which has potentially large impact on the extent and method (hence cost) of remedial construction required to assure adequate seismic performance of large dams. For example, it may reduce the cyclic shear stress ratio predicted to cause liquefaction in a soil layer under a typical large dam (of the order of 30m high) to about 50% of its value in the absence of the dam. The predicted deformations and resistance and the remedial strength required are a direct function of the reduced shear strength. The effects of the K_σ strength reduction is driving the majority of seismic dam safety concerns and needs for remediation, which may result in costs in the billions of dollars in North America alone in the coming years.

The factor K_σ quantifies the curvature in the cyclic shear strength envelope (cyclic shear stress required to cause liquefaction versus confining stress) for a soil as observed in laboratory tests on discrete specimens. Although some curvature may be expected, such large reductions in cyclic shear strength ratios are counter-intuitive. It is generally accepted that increased confining stress should broadly improve the capacity of a soil to resist applied loads, not reduce it. Clearly the volume of soil which is required to be treated and the difficulty and expense of that treatment are highly dependent on an accurate assessment of the potential for and consequences of liquefaction.

There is therefore strong motivation for owners of large dams to investigate the behavior of saturated sands subject to strong ground shaking under high initial effective confining stresses. In the absence of field data, the use of a centrifuge was considered to be the only practical option to realistically represent a deep soil deposit subjected to earthquake shaking. The studies to date have focused on level ground initial stress conditions in a two-layer (dense over loose) deposit of clean, fine Nevada sand. Examination of more complex stratigraphy and sloping ground stress conditions is planned for future studies.

2.0   The ERDC Centrifuge, earthquake actuator and ESB specimen container

The design specification for the ERDC centrifuge followed a review of the Army's research needs and a study of the available academic facilities, Ledbetter (1991). Many of the field problems with which the U.S. Army Corps of Engineers is concerned are physically large, such as earth dams, locks and river control structures, environmental problems and military research. It was determined that a new facility with a high payload capacity and high g capability was required to meet future Army needs. A large beam centrifuge was commissioned, termed the Acutronic 684, based on the French designed Acutronic 661, 665 and 680 series of geotechnical centrifuges, Ledbetter et al. (1994a). The capacity of the ERDC centrifuge is a payload of 8 tonnes at up to 143 g, reducing to 2 tonnes at 350 g, with a platform area of 1.3m square. This high capacity enables field problems of the order of up to 300m in breadth, 300m in depth, and 1000m in length to be simulated under a wide variety of loading conditions. The facility is now fully operational, and is equipped with a large range of equipment and appurtenances, Figure 1.

Figure 1. The WES centrifuge

The research approach for this high confining stress liquefaction study uses the large capacity of the WES centrifuge to investigate the generation of excess pore pressure and liquefaction under conditions which much more closely resemble those at depth in the field, Ledbetter et al. (1999). To do this required the design and construction of a large dynamic actuator.

2.1   Shaker design

A mechanical design was adopted for the earthquake actuator for the ERDC centrifuge, based closely on a smaller version designed for Cambridge University, England. Complex servo-hydraulic shaking systems have been developed and are operational on several centrifuges around the world, but these are limited in the g level to which they may operate, and are considerably more expensive to build. The research by ERDC into the basic behavior of liquefying soils and comparison with present design methods required a base motion comprising a series of uniform cycles of base shaking, of variable duration and uniform frequency. A rotating mechanical system is ideally suited for this purpose. The large carrying capacity of the centrifuge meant that the shaker itself could act as the reaction mass for the specimen, minimizing the vibrations which would be otherwise transmitted to the centrifuge itself, and the model container could be much deeper and longer than usual.

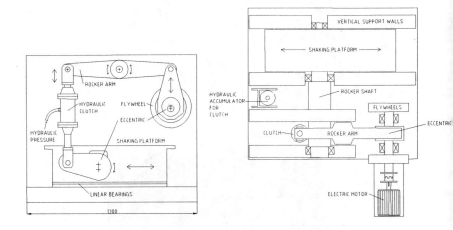

Figure 2. The ERDC Mk I earthquake actuator

The principle of the operation of the Mk I shaker was based on the use of stored energy to drive the specimen back and forth. Flywheels are incorporated in the shaker mechanism to store the energy in advance of the shaking event in the form of stored angular momentum. A system of linkages and eccentrics transferred the stored energy of the fly wheels to the shaking platform and thence into the soil specimen, Figure 2. A hydraulic or electrical motor drove the flywheels up to full speed, and then, on a signal, a high speed clutch grabbed the oscillating shaft and transferred energy into the model until another signal released it again. Clearly the frequency of the oscillation was directly proportional to the speed of the motor (and flywheels). The amplitude was controlled by the arrangement of the eccentrics; three displacement amplitudes for the platform were available (+/- 0.49mm, +/-1.47mm and +/-4.41mm).

The shaker was designed (structurally) for operation up to 150g, at which the maximum load capacity of the shaking platform is reached (75 tonnes). The design maximum lateral force which the mechanism could exert on the shaking platform was 30 tonnes, and the maximum frequency at which the shaker could be safely operated was 150 Hz (eg. 1Hz prototype at 150g, or 3Hz prototype at 50g). (The Mk I shaker has recently been upgraded.)

2.2 Model containment

The specimen is built within a hollow rectangular model container, termed an equivalent shear beam (ESB) container, comprising a series of eleven aluminium alloy rings stacked one above the other, and separated by an elastic medium, Figure 3. Several of these chambers have been constructed, and extensive dynamic analysis and testing has been carried out to determine their dynamic response characteristics, Butler (1999). The model container has internal dimensions of 627mm deep by 315mm wide by 796mm long. Each of the eleven aluminium alloy rings is 50mm high. The rings are not stiff enough along their long dimension to support the outward pressure from the soil inside under high g, but they are supported by the massive reaction walls of the shaker unit itself. A rubber sheet separates the rings from the steel walls on either side. This concept has the added advantage of raising the center of gravity of the reaction mass in line with the center of gravity of the specimen, thus minimizing eccentric forces that may lead to rocking.

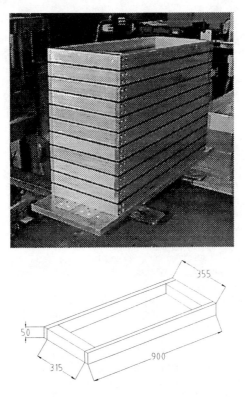

Figure 3. Typical Equivalent Shear Beam specimen container

Thin metal sheets, termed shear sheets, are positioned on the interior end walls of the chamber and fixed securely to its base. The shear sheets accommodate the complementary shear force generated by the horizontal shaking within the specimen and transmit that force to the base of the container, Figure 4. This improves the uniformity of the stress field at each elevation along the model, reducing the tendency for the chamber to 'rock'.

Figure 4. Shear sheets form the boundaries on the end walls of the ESB

The ESB concept is to create an equivalent shear beam with an average stiffness comparable to the stiffness of the soil specimen. Expressed rigorously, the concept is more accurately defined as achieving a dynamic response that does not significantly influence the behavior of the soil specimen inside. In certain classes of experiment, where the zone of interest is limited to the central region of the chamber, it may be expected that the stiffness of the soil (at least near the end walls) would not reduce significantly during shaking due to excess pore pressure rise. In this case the stiffness of the chamber may be designed accordingly, perhaps considering a shear modulus appropriate to the level of dynamic strain expected in the soil free-field at mid-depth. For other experiments involving the liquefaction of large volumes of soil inside the container, the stiffness changes (and hence dynamic response) will change dramatically throughout the base shaking. A stiff chamber may lead to undesirable effects, as noted by Peiris (1999) who observed that liquefaction in a loose saturated sand model did not occur near the stiff end walls of the chamber. A chamber with no stiffness simply adds mass to the soil specimen, again changing its dynamic response. This poses a particular challenge for the design of an ESB container.

The ERDC ESB used in these experiments was assembled using a urethane adhesive sealant (commonly used as a windshield sealant for cars) between the aluminum alloy rings. This material has good elastic properties (exhibiting only minimal hysteresis under cyclic loading) and bonded well to the metal and to itself. The ESB has a relatively low shear stiffness of 441 KN/m^2 (shear stiffness of the full stacked ring assembly) and mass of 229 kg, with a first mode at 16Hz, and second, third, and fourth modes at 46, 87, and 116Hz respectively (Butler 1999). A typical saturated specimen at 50g in the ERDC ESB will have a theoretical natural frequency of around 84Hz, based on an average small strain shear modulus of 96 MN/m^2.

2.3 Dynamic response of the model container and specimen

In his doctoral thesis, Butler (1999) completed a thorough theoretical and experimental analysis of the dynamic response of the coupled soil-container system. At high g the soil and container act as a coupled system, where the lower stiffness of the container reduces the natural frequency (slightly) of the combined system compared to the soil column alone. However, provided the driving frequency is low relative to the natural frequency of the coupled system, Butler (1999) demonstrates that the displacement response of the system is unaffected compared to the soil acting independently, an ideal situation.

For higher driving frequencies, Butler (1999) concludes that it would be necessary to reconsider the elastic stiffness of the ESB container, and to tune the container to ensure that even with the expected level of degradation in the soil specimen, the coupled system did not deviate significantly from the condition of the soil column alone. This may be possible by adding mass to the rings of an initially stiffer ESB to reduce its first mode to the desired level, Butler (1999).

The liquefaction of a level sand bed has previously been the subject of other research. Experiments were conducted at many centrifuge centers under the VELACS project, Arulanandan and Scott (1994). The objective of the ERDC study is to investigate the onset of liquefaction under much higher initial effective overburden stresses.

Model series	Models in series	Effective overburden stress in loose layer	Depth of prototype (approx)	Depth of specimen	Notes
2	a, b, c, d, e, f	1 tsf	15 m	300 mm	Nevada sand
3	a, b, c, d, e	2 tsf	26 m	525 mm	Nevada sand
4	a, b, c, d	3 – 5 tsf	26 – 40 m	525 mm	Nevada sand with lowered w.t. or surcharge
5	a, b, c, d	7 – 10 tsf	54 – 63 m	525 mm	Nevada sand with lead surcharge

Table 1. Summary of model tests

3.0 Research Program

Table 1 summarises the experiments conducted during 1998 and 1999. The models are grouped in series, where each series corresponds to a different target range of vertical effective overburden stress in the loose layer. In all cases, the bottom

160mm of the specimen was around 50% Relative Density (RD) and the upper portion was around 75% RD. All models were shaken at 50g. Some models were overconsolidated by a factor of 2.5 prior to shaking (achieved by running the centrifuge up to 125g). A large series of experiments have been conducted with a range of overburden depths to provide repeatability and redundancy in the instrumentation and data records. All the models were built using Nevada sand and tested at 50 g. The experiments are summarized in Table 2.

Model Code	Overall depth (mm)	Relative Density	σ_v' at mid-depth in loose layer (tsf)	OCR	Number of earth-quakes	Comments
2a	300	44% loose 83% dense	1	1	3	Saturated to ground surface.
2b	300	50% loose 75% dense	1	1	2	Saturated to ground surface.
2c	300	49% loose 74% dense	1	1	5	Saturated to ground surface.
2d	300	50% loose 75% dense	1	1	4	Saturated to ground surface.
2e	300	49% loose 73% dense	1	2.5	4	Saturated to ground surface.
2f	300	50% loose 75% dense	1	2.5	4	Saturated to ground surface.
3a	525	34% loose 73% dense	2	1	2	Saturated to ground surface.
3b	525	49% loose 77% dense	2	1	3	Saturated to ground surface.
3c	525	49% loose 79% dense	2	1	3	Saturated to ground surface.
3d	525	54% loose 80% dense	2	2.5	4	Saturated to ground surface.
4a	525	49% loose 80% dense	3	1	4	Saturated to top of loose layer only.
4b	525	56% loose 74% dense	3	2.5	4	Saturated to top of loose layer only.
4c	525	50% loose 75% dense	4.7	1	4	Saturated to ground surface. Lead surcharge.
4d	525	50% loose 68% dense	4.7	2.5	4	Saturated to ground surface. Lead surcharge.
5a	525	51% loose 72% dense	7.4	1	4	Saturated to ground surface. Lead surcharge.
5b	525	49% loose 76% dense	7.4	2.5	4	Saturated to ground surface. Lead surcharge.
5c	525	52% loose 75% dense	9.2	1	3	Saturated to ground surface. Lead surcharge.
5d	525	57% loose 80% dense	9.2	1	1	Saturated to ground surface. Lead surcharge.

Table 2. Detailed summary of experiments

The Nevada sand used in the models was characterized by standard laboratory tests to determine parameters such as dry density and gradation. Table 3 presents key material parameters for this sand.

Specific gravity	2.64
Maximum void ratio	0.757 (density 93.8 pcf)
Minimum void ratio	0.516 (density 108.7 pcf)
D_{50}	0.18 mm (approx)
D_{10}	0.11 mm (approx)

Table 3. Nevada Sand (parameters as measured)

The pore fluid comprised a mixture of glycerine and water, 80% by weight for experiments conducted at 50g. Measurements of the viscosity of glycerine-water mixes at a range of temperatures and proportions show that the viscosity is sensitive to both parameters.

The density of a glycerine-water mix was calculated from:

$\rho_m = \rho_g(m_g + m_w)/(m_g + \rho_g m_w)$

where ρ_m is the density of the mix, ρ_g is the density of glycerine, m_g is the mass of glycerine, and m_w is the mass of water. Table 4. summarizes the properties of the glycerine-water solution used as the pore fluid.

Density	1200 kg/m^3
Viscosity	50 cs
Specific Gravity	1.26
Composition	80% glycerine-water mix (by weight)

Table 4. Parameters for pore fluid (as measured)

The models were poured dry from a hopper and saturated under vacuum, or slowly under gravity. Instrumentation was placed in the model as it was being constructed.

The typical arrangement of the model specimens for the different test series is shown in Figure 5. Instrumentation was positioned through the depth of the models, and comprised pore pressure transducers and accelerometers. The location of the instrumentation for Model 5b is shown in Figure 6.

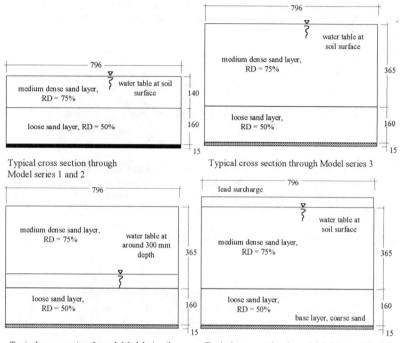

Typical cross section through
Model series 1 and 2

Typical cross section through Model series 3

Typical cross section through Models 4a, 4b

Typical cross section through Models 4c, 4d and 5

Figure 5. Cross sections through the different model configurations

4.0   Amplification of base input motion

As the base input motion propagates through the height of the specimen, it may be amplified or attenuated depending on the dynamic properties of the soil column and the frequency content of the input motion. This is similar to conditions in the field, where elastic wave energy is modulated as it passes through different soil deposits. Examination of the time histories of acceleration at different depths in a deep soil column such as Model 5b, Figure 6, shows that in the absence of significant degradation caused by large strains or excess pore pressures, the amplification factor is around 1, Figure 7. A similar response was also found for shallow specimens. In specimens where large excess pore pressures started to develop through the soil column, however, the characteristics of the motion propagating upwards tended to change dramatically. In the extreme, the lateral accelerations of the surface becomes isolated from the base input, as may be seen in Figure 8 which shows the acceleration at different depths in Model 2f.

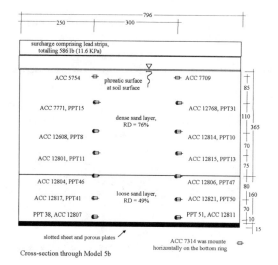

Figure 6. Instrumentation for Model 5b

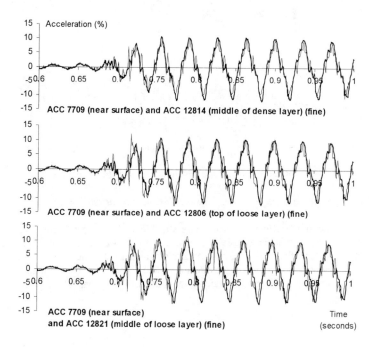

Figure 7. Comparison of the measured surface acceleration with the acceleration at different depths, Model 5b

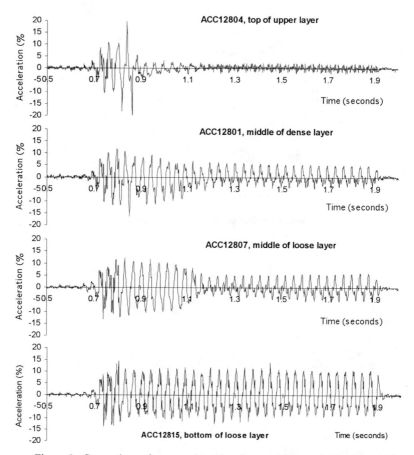

Figure 8. Comparison of measured accelerations at different depths, Model 2f

5.0   *Development of excess pore pressures and liquefaction at shallow depths*

The target initial effective vertical stress in Model 2f was 1 tsf (100 KPa) at mid-depth in the loose layer (Table 2). The specimen was overconsolidated by accelerating the centrifuge to 125g prior to subjecting it to shaking motion at 50g. Figure 9 shows how the excess pore pressures at different depths developed with time. It is clear that within a few cycles of shaking, the loose sand layer has fully liquefied.

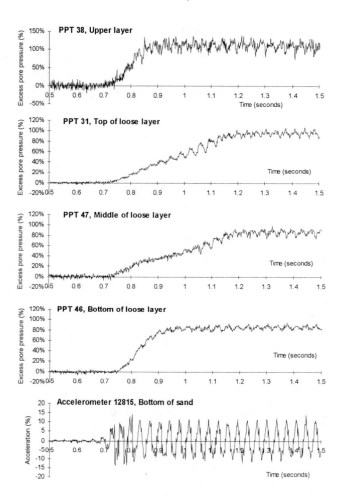

Figure 9. Development of excess pore pressure in a shallow bed, Model 2f

Comparison between the acceleration time histories in Figure 8 and the excess pore pressures in Figure 9 for the loose and dense layers shows the correlation between transmitted motion and level of excess pore pressure. The upper section (near surface) of the dense layer liquefies within 0.2 seconds (0.2 x 50 = 10 seconds field equivalent). The percentage of excess pore pressure is calculated by dividing the measured fluid pressure by the calculated initial overburden stress. The actual overburden stress may vary over time due to small movements of the transducer in the liquefied ground; this has a larger effect on the accuracy of measurement at shallower locations, such as PPT 38.

It is interesting to note that despite the difference in relative density between the dense and loose layers, the soil column behaves more like a single unit than as two layers. This is because each increment of excess pore pressure at depth immediately causes a fluid pressure, which affects the whole column through the volumetric compressibility of the pore fluid. This is considerably lower than the shear compressibility of the soil skeleton, which is determining the speed of transmission of the elastic shear wave upwards.

6.0   Generation of excess pore pressures at high confining stress

At high initial effective confining stress pore pressures are also observed to rise strongly during the initial cycles of shaking. After some time, this steady rise is arrested, and then excess pore pressure is capped at a value less than the full 100% of the initial vertical effective stress seen in shallow soil columns. Figure 10 shows a set of data from Model 4d, a deep soil specimen with an initial effective vertical stress of around 500 KPa (4.7 tsf) at mid-depth in the loose layer.

A similar picture was found at even higher initial vertical effective stress. Figure 11 shows the development of excess pore pressure under a vertical effective stress in the middle of the loose layer of 1037 KPa (9.7 tsf). In this figure, the actual fluid pressure has been plotted in KPa, to be compared with the initial vertical effective stress. The time history of base shaking is shown below the pore pressure record in two forms. The first trace shows the measured record, which includes a degree of high frequency noise. In the second, lower trace, this signal has been processed with a Butterworth band pass filter, 250 and 10 Hz, to extract the main shaking frequency. From this record, it is clear that the amplitude of the shaking motion is almost constant throughout the record, and particularly that the amplitude remains at or near its peak level throughout the second half of the shaking episode, when the excess pore pressure has been capped (after about 0.8 seconds of shaking, or 40 seconds field equivalent). In this case, the excess pore pressure is capped at around 50% of the value required to reach 'initial liquefaction'.

The amplitude of cyclic shaking was easily sufficient to generate excess pore pressures at a strong rate during the first half of the event, and this can be seen in the top trace of Figure 11. Indeed, after around 0.4 seconds of shaking the rate of excess pore pressure build-up increases, and this may be related to a moderate increase in the amplitude of base shaking at the same time. Between around 0.4 seconds and 0.7 seconds of shaking the rate of increase of excess pore pressure is around 350/0.3 ~ 1000 KPa/second (or 20 KPa/second field equivalent). The driving frequency of input motion is around 50 Hz (1 Hz field equivalent). During the same period, the amplitude of shaking is around +/- 15% g, which corresponds to a cyclic displacement of +/- 0.76 mm (38 mm field equivalent) at that depth.

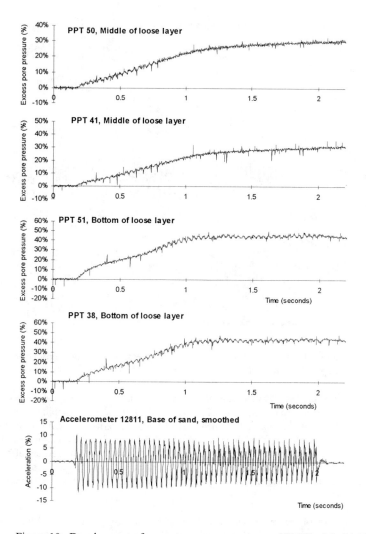

Figure 10. Development of excess pore pressure at over 500 KPa, Model 4d

This pattern of behavior was found in all specimens at initial vertical effective confining stresses of around 250 KPa or greater. Although not a rigorous comparison, as there are a number of variables (including amplitude) between the different experiments, Figure 12 shows the normalised results from a range of initial conditions. There is a general trend with increasing initial vertical effective stress, towards the capping of excess pore pressure at a level lower than the expected 100%.

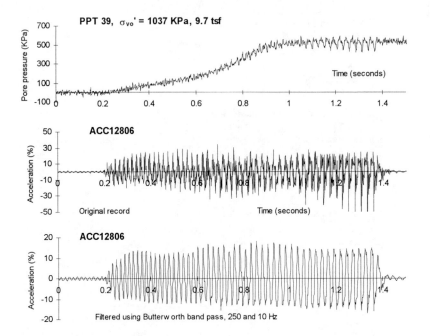

Figure 11. Development of excess pore pressure at initial vertical effective stresses of up to 1000 KPa (10 tsf), Model 5d

All of the records of excess pore pressure from all the model specimens were re-examined to determine the residual pore pressure at the level of capping. Although the amplitudes of shaking and the densities of the sand varied (the data-set includes both 'dense' and 'loose' sand layers as described above) a clear correlation was found between the maximum level of excess pore pressure and the initial vertical effective stress, as shown in Figure 13. This figure includes 125 data points from first and succeeding earthquakes, normally and overconsolidated specimens, and dense and loose layers.

Figure 14 shows the data from first earthquakes only, broken out by normally consolidated and overconsolidated (including dense and loose results). Again, there is a strong correlation between the maximum level of excess pore pressure generated during the shaking and the initial vertical effective stress. In this case, the overconsolidated data is seen to lie to the left of the normally consolidated data, indicating that the level of excess pore pressure may reduce for increased overconsolidation. This is consistent with experience.

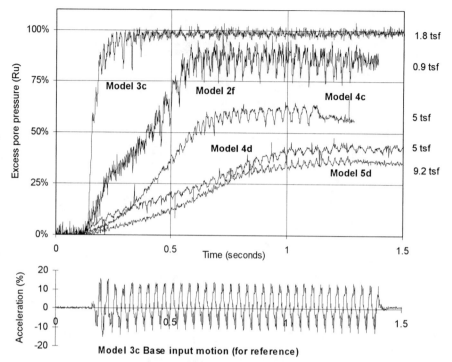

Figure 12. Comparison between excess pore pressure development at different initial vertical effective stresses

On the right hand side of Figures 13 and 14, an approximate depth scale has been given for guidance, expressed in feet. Clearly the calculation of equivalent field depth depends on many assumptions, but it is worth noting that the data appear to show that whilst liquefaction may readily occur within the upper 40 feet or so of a saturated sand deposit, it becomes increasingly unlikely at greater depths. This is consistent with field observation.

Research work is continuing with a view to full verification of these findings and the preparation of design guidance. Further experiments will address the effects of amplitude of shaking, sloping ground and density.

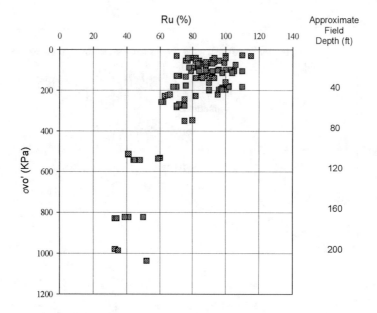

Figure 13. Comparison between maximum residual excess pore pressure and initial vertical effective stress, all data

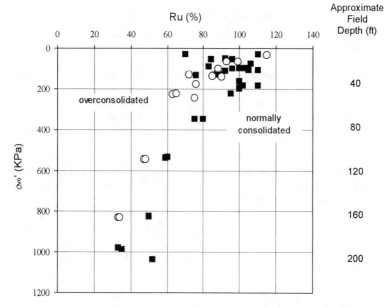

Figure 14. Comparison between overconsolidated and normally consolidated data, all data

7.0 Consequences for a hypothetical large earth dam

To illustrate the potentially highly significant consequences of capped pore pressures decreasing with depth, a hypothetical earth dam of the form outlined in Figure 15 is assessed. The estimated deformation response is calculated for earthquake induced potential pore pressures based on standard approaches, Seed and Harder (1990), Youd and Idriss (1997), and on limiting pore pressure generation. Deformations are estimated following the procedures of Ledbetter and Finn (1993), Finn et. al. (1995), and Ledbetter et. al. (1994b), with the use of the non-linear dynamic effective stress analysis program TARA, Finn el. al. (1986), and Finn and Yogendrakumar (1989).

The dam is located across stream channel deposits with assumed profile, Figure 15, as: (a) foundation older stiffer alluvium below elevations 151 m (u/s) - 163 m (d/s), (b) younger less stiff alluvium above the older alluvium, and (c) roller compacted zoned earth dam founded on the young alluvium. Crest level and Pool level are 211m and 199m respectively. Standard Penetration $(N_1)_{60}$ values range from 5 to 15 and shear-wave-velocity (V_s) from 180-450 m/s in the young alluvium freefield upstream and downstream. Beneath the dam, the young alluvium downstream $(N_1)_{60}$ values range from 30 – 40 and V_s from 425-700 m/s and upstream $(N_1)_{60}$ values range from 6-20 with V_s from 150-350 m/s. Ground motions are assumed to be from a magnitude 7.75 earthquake occurring at a distance from the dam about 50 miles with a peak acceleration of 0.22g at the dam.

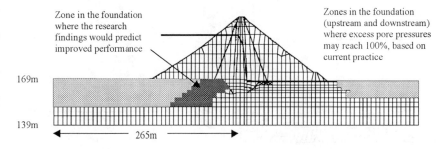

Figure 15. Comparison of affected zones in the foundation

Maximum potential residual excess pore pressure

Seed's simplified procedures (Seed and Harder, 1990) were coupled with non-linear dynamic effective stress finite element analysis to yield "field observation and laboratory based" estimates for earthquake induced pore pressures. These pore pressures are considered the "maximum potential residual excess pore pressures" that that could be generated by the design earthquake, based on historic field earthquake response. Unless the soils have low permeability and/or are bounded by low permeability layers that prevent the dissipation of excess porewater pressures, earthquake-generated porewater pressures should be less than the maximum potential.

For this example dam, Figure 15 shows the maximum potential residual excess porewater pressures (shaded zones) that can be generated by the design earthquake. The young alluvium can generate pore pressures (a) downstream of the downstream toe but nothing of significance beneath the downstream shell and (b) upstream of the upstream toe and beneath the upstream shell. The lighter shaded upstream zone extends completely beneath the darker shaded zone. Pore pressures in these lighter shaded zones are between 80 – 100 % of the initial vertical effective stress with the majority between 90 and 100 %.

The research findings expressed in this paper are for level ground conditions. Continuing research will be addressing the effects of residual shear stresses such as induced by the geometry of a dam. Until the geometry effects are known, this example applies the pore pressures trend shown in Figure 14 for the normally consolidated case as the cap or limitation. In this example, the zone of soil that will be effected and have reduced pore pressure potential is the darker shaded zone beneath the upstream shell; quite naturally, this is a high confining stress zone and where K_σ leads to a strength reduction due to the high confinement. This zone is exactly where strength increase or less strength degradation is of prime importance to the dam behavior under earthquake loading. The pore pressure potential in the darker shaded zone is reduced from 80 – 100 % to 50 –75% of the vertical effective stress.

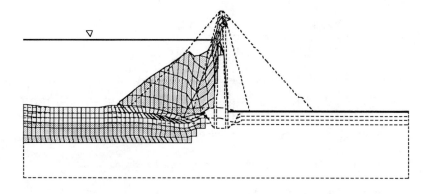

Figure 16. Deformations calculated using current analyses

Deformation response of the dam

Figures 16 and 17 illustrate the effects or differences in earthquake induced deformations and strains caused by the applied pore pressure limits. For the deformation stage shown in the figures, the only elements that appear shaded are those significantly straining and participating. Figure 16 shows the deformations at effective strength reduction to 72 % of the initial caused by pore pressure increase. The dam will continue to deform as pore pressure builds to the maximum potentials and residual strength is reached beneath the dam in the foundation, which is around 90 % reduction of initial strength. Unless remediated, severe earthquake damage to the dam is possible accompanied by overtopping and piping through the core resulting in complete failure.

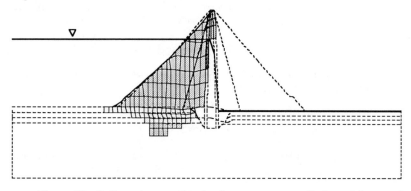

Figure 17. Deformations predicted using pore pressure limits and improved soil strength

Figure 17 illustrates the benefits achieved by applying the limiting pore pressure with depth trend. The deformation stage in Figure 17 is the same as that in Figure 16. As shown strains, deformations and the active elements are significantly reduced. The dam continues to deform as pore pressures increase to their maximum potentials and residual strengths; however, the final stages of deformation are considerably less than those under the conditions in Figure 16.

Immediately obvious from comparison of Figures 16 and 17 is that more strength is available to resist deformation movements and to buttress and assist remedial treatments. The availability of more strength translates to smaller and less remediation and lower costs. The stage of progressive strength reduction illustrated in both Figures 16 and 17 was chosen to best exemplify the impact of limiting pore pressure criteria on dams with liquefiable soils in high initial effective confining stress fields.

The implication of the present study is that although soil at depth may soften and degrade under earthquake induced excess pore pressure behavior, the reduction in strength is limited and liquefaction under high confining stresses may not be the hazard that it is presently assumed to be.

8.0   Conclusions

1.   A large data-set of the behavior of loose saturated sands under high initial effective confining stresses and subject to earthquake-like shaking has been collected during an extensive experimental program on the ERDC Centrifuge, Vicksburg MS.

2.   This data has shown that, at moderate amplitudes of excitation, the maximum level of excess pore pressure development is capped at high initial effective confining stress and does not reach a level sufficient to cause 'initial liquefaction' (defined as 100% of the initial vertical effective stress).

3.   These findings are currently being verified and will then be used to develop appropriate design guidance.

4.   High pore pressures and the potential for liquefaction beneath dams may not be the hazard that it is currently perceived to be.

5.   The implication for the assessment of liquefaction hazard and requirements for remediation works under large earth dams is potentially very significant.

Acknowledgements

The ERDC Centrifuge Research Center and the Army Civil Works Earthquake Engineering Research Program of the U.S. Army Corps of Engineers sponsored this research. The members of the ERDC centrifuge team who helped conduct this work were Richard Burrows, Lee Miller, Dr Gary Butler, Rodgers Coffing, Wipawi Vanadit-Ellis, and Dr Mike Sharp. The ERDC machinists contributed greatly to the project. Drs William Marcuson and Michel O'Connor were Directors of the Geotechnical Laboratory. Permission to publish this paper by the Chief of Engineers is gratefully acknowledged.

References

Arulanandan, A. and Scott, R.F. (1994) editors, Proc. Int. Conf. on Verification of Numerical Procedures for the Analysis of Soil Liquefaction Problems, at U.C. Davis; Volumes 1 & 2; Balkema.

Butler, G.D. (1999). A dynamic analysis of the stored energy angular momentum actuator used with the equivalent shear beam container, PhD thesis, Cambridge University (in preparation).

Finn, W.D.L. and Yogendrakumar, M., Yoshida, N. and Yoshida, H., (1986), "TARA-3: A Program to Compute the Response of 2-D Embankments and Soil-Structure Interaction Systems to Seismic Loadings", Department of Civil Engineering University of British Columbia, Canada.

Finn, W.D.L., Ledbetter, R.H., and Marcuson, W.F. III (1995). "Seismic Deformations in Embankments and Slopes", Symposium on Developments in Geotechnical Engineering - From Harvard to New Delhi, 1936-1994, Bangkok, Thailand, A.A. Balkema, Rotterdam.

Finn, W.D.L. and Yogendrakumar, M., (1989) "TARA-3FL; Program for Analysis of Liquefaction Induced Flow Deformations", Department of Civil Engineering, University of British Columbia, Vancouver, Canada.

Ledbetter, R.H. (ed) (1991) Large centrifuge: a critical Army capability for the future, Misc. Paper GL-91-12, Dept. of the Army, Waterways Experiment Station, Vicksburg, MS, May.

Ledbetter, R.H., and Finn, W.D.L. (1993). "Development and Evaluation of Remediation Strategies by Deformation Analysis", ASCE Specialty Conference on Geotechnical Practice In Dam Rehabilitation, Raleigh, North Carolina, April, pp 386 - 401.

Ledbetter, R.H., Steedman, R.S., Schofield, A.N., Corte J.F., Perdriat J., Nicholas-Font J. and Voss H.M. (1994a) US Army's engineering centrifuge: Design, Proc. Centrifuge 94, pp63-68, Singapore, 31 Aug – 2 Sept, Balkema.

Ledbetter, R.H., Finn, W.D.L., Hynes, M.E., Nickell, J.S., Allen, M.G., and Stevens, M.G. (1994b). "Seismic Safety Improvement of Mormon Island Auxiliary Dam", 18th International Congress on Large Dams, Durban, South Africa.

Ledbetter, R.H., Steedman, R.S. and Butler, G.D. (1999) Investigations on the behavior of liquefying soils, Proc Int Workshop on Physics and Mechancs of Soil Liquefaction, Baltimore MD, Lade P.V. and Yamamuro J.A. (eds), 10-11 Sept 1998, Balkema, pp 295-306.

Peiris, L.M.N. (1999) Seismic modelling of rockfill embankments on deep loose saturated sand deposits, PhD Thesis, Cambridge University.

Seed H.B. (1983) Earthquake-Resistant Design of Earth Dams, Proceedings of Symposium on Seismic Design of Embankments and Caverns, May, ASCE pp 41-64.

Seed R.B. and Harder L.F. (1990) SPT-Based analysis of cyclic pore pressure generation and undrained residual strength, Proc. H Bolton Seed Memorial Symposium, Vol. 2, pp 351 - 376, BiTech Publishers Ltd, Vancouver.

Youd, L. and Idriss, I., Eds (1997) Workshop on Evaluation of Liquefaction Resistance of Soils, Proceedings, Salt Lake City, Technical Report NCEER-97-0022, sponsored by FHWA, NSF and WES, published by NCEER.

Liquefaction of Cohesive Soils

Vlad G. Perlea[1], Member, ASCE

Abstract

Liquefaction of sand, clean or with some fines content, has been extensively studied over the last three decades and is currently a phenomenon reasonably predictable. The study of cohesive soils behavior during and immediately after cyclic loading is especially difficult because of their structure variability and the major influence of structural characteristics on dynamic properties, which makes testing of reconstituted samples of little interest in evaluation of natural deposits. There are several categories of cohesive soils potentially liquefiable, which are considered in this paper: clayey silts or silty clays of low plasticity meeting the "Chinese Criteria"; highly sensitive clays; collapsible loess. Data from literature for better understanding of their susceptibility to liquefaction (triggering, post-cyclic strength) are summarized.

Introduction

The liquefaction of relatively loose saturated sandy soils under cyclic loading is attributed to the increase in pore pressure due to the tendencies of the soil particles to re-arrange into a denser state. The most disturbing effect of sandy soil liquefaction, i.e. its significant loss of shear strength, is a direct consequence of the release of contacts between particles following the decrease of the effective confining pressure. Although cyclic undrained loading would increase the pore pressure in cohesive soils as well, their "cohesion" prevents separation of particles and, therefore, the loss of shear strength is generally less dramatic than in the cohesionless soil case. However, shear deformation induces re-arrangement of particles and may result in structural collapse and corresponding pore pressure build-up and severe loss of shear strength, both undrained and drained.

----

[1]Civil Engineer, Geotechnical Branch, U.S. Army Corps of Engineers, Kansas City District, 601 East 12th Street, Kansas City, MO 64106-2896

There are various loading conditions that can generate liquefaction-like loss of shear strength in cohesive soils, including static (monotonic increasing) loading. However, because of the major interest in stability evaluation of natural deposits during earthquakes, only seismic type loading is considered in this paper. There are three main items of interest for behavior evaluation: the level of loading which may trigger strength loss, evolution of strength degradation during seismic loading, and post-earthquake strength. In this report, cohesive soils are considered natural materials with at least 50% fines (particles passing the 75-μm sieve), any measurable liquid limit (LL), and the plasticity index (PI) at least 4.

Case Histories

During strong earthquakes in China from 1966 to 1976, especially during Haicheng (1975) and Tangshan (1976) earthquakes, many cohesive soil deposits liquefied, as observed through ejection of liquefied soil to surface. According to Wang (1979) the cohesive soils liquefied during these events had: less than 20% clay fraction, liquid limit between 21 and 35, plasticity index 4 to 14, and water content higher than nine-tenths of the liquid limit.

Liquefaction of fine-grained soils (defined in ASTM D 2487 as having 50% or more fines, i.e. particles passing the 75-m sieve) have also been observed in Japan. For example: Kishida (1969) reported liquefaction of soils with up to 70% fines and 10% clay fraction; Tohno and Yasuda (1981) reported liquefaction of soils with up to 90% fines and 18% clay due to the 1968, Tokachi-Oki earthquake; Miura et al. (1995) noted liquefaction of soils with up to 48% fines and 18% clay due to the 1993, Hokkaido Nansai-Oki earthquake. Based on studies by H. Tsuchida of liquefactions in 1964 in Niigata, Japanese norms consider easily liquefiable soils with up to 100% fines and up to 25% clay fraction (Japan Society of Civil Engineers, 1977).

In the United States, the upstream flow slide failure of the Lower San Fernando Dam, consequent to the 1971 San Fernando earthquake, has been attributed to liquefaction of "very silty" hydraulic fill sands (Seed et al., 1989). Liquefaction in soils containing silty or clayey fines is likely, though not well documented, to have occurred as a result of the 1886 Charleston, South Carolina earthquake (Obermeier et al., 1985). Alternating layers of silty and/or clayey glaciomarine sand deposits have been contended to have liquefied during several northeastern USA earthquakes (Tuttle and Seeber, 1989). Wesnousky et al. (1989) detail a comprehensive evaluation of geologic evidence of extensive liquefaction and massive sliding of loessial bluff soils throughout the Mississippi embayment as a result of 1811-12 New Madrid, Missouri series of very strong earthquakes. Youd et al. (1985) reported liquefaction of soil with 70% fines and 20% clay fraction (< 5 μm) at the Whiskey Springs site (2 km from epicenter) during the M = 7.3 Borah Peak, Idaho earthquake.

Preliminary investigation of the effects of the recent Izmit earthquake in Turkey (August 1999, local magnitude 7.4) showed that buildings in the city of Adapazari (30 km epicentral distance) rotated as much as 50 degrees or sank to a depth as great as 2 meters due to liquefaction of silts and sand-silt-clay soil mixtures (Hynes, 1999).

Artificial materials deposited in tailing dams or lagoons are also susceptible to liquefaction. Ishihara (1985) found that lean clay from the El Cobre, Chile, No.4 dike (copper tailings slimes, 95% fines, 20% clay fraction, PI = 11), failed by liquefaction during the 1964 earthquake in Chile, has the resistance to liquefaction comparable with reconstituted samples of clean sand with a relative density of 40%. Gold mine tailings (silt, LL = 31%, PI = 10, in-situ water content = 37%) liquefied at the time of Izu-Oshima-Kinkai in 1978 in Japan (Ishihara, 1984).

A particular category of cohesive soils susceptible to liquefy under seismic action are deposits already on the verge of failure before the earthquake occurrence. Ishihara et al. (1990) reported that an earthquake of magnitude 5.5 shook the Dushanbe region of Tajikistan in 1989, whereupon wetted, low-plasticity loess (80% silt, 15% clay, LL = 30, PI = 10, w = 40%) slumped and flowed as far as 2 km. However, the authors state that *"this event seems to be unique and previously unknown phenomenon"* and *"the loess deposit between the depths of about 7 and 17 m appear to have been in a state of impending hydraulic collapse even prior to the advent of the earthquake"*. In Bulgaria, the 1977 Vrancea earthquake triggered landslides in the saturated loess of the Danube river bank, at approximately 180 km epicentral distance. The catastrophic liquefaction of dry loess in Kansu Province, China during a strong earthquake in December 1920 destroyed over 100,000 lives. Air pressure generated by the collapse of cave homes induced tens of avalanches of dry dirt, burying cities, villages, farmsteads, and stream beds (Hall and McCormick, 1922). The collapsible Japanese "Shirasu", similar to loess in gradation and structure, is also liquefiable.

Liquefaction induced landslides occurred in 1964 in Anchorage, Alaska (Seed and Wilson, 1967). A 1.5 to 6 m surface layer of sand and gravel was underlaid by a deep bed of Bootlegger Cove clay, a sensitive deposit of marine clay; the most sensitive material had typically LL = 33 and PI = 12.

Criteria for Identification of Cohesive Soils Potentially Liquefiable

Based on observations during Chinese earthquakes, Wang (1981) identifies three categories of liquefiable soils: (1) saturated sand, with Standard Penetration Test (SPT) blow counts lower than a critical value, function of earthquake intensity and effective overburden pressure; (2) saturated slightly cohesive silty soils, with the water content lower than 90% of its liquid limit and the liquidity index greater than 0.75; and (3) soft clays with the liquidity index greater than 0.75, the unconfined compressive strength less than 0.5 daN/cm^2, SPT blow counts less than 4, and sensitivity in excess of 4.

The first reputable reference in the American technical literature on liquefiability of some cohesive soils is due to Seed et al. (1983): *"Consider some clay soils as being vulnerable to significant losses in strength. Based on Chinese data, these soils would*

appear to have the following characteristics: percent finer than 0.005 mm < 15%, liquid limit, LL < 35, and water content > 0.9 LL. The best way to handle these soils, if they plot above the A-line, would be to determine their liquefaction characteristics by tests."
However, the Chinese practice of determining clay fraction, plastic limit, liquid limit, and moisture content differs from procedures used in the USA and many other countries. The most notable difference is the use of fall-cone penetrometer for plasticity limits in People Republic of China (PRC Soil Testing Standard SD 128-007-84) as compared with percution test in the Casagrande device and rod-rolling method in USA (ASTM D 4318). Based on studies performed at Vicksburg District of the US Army Corps of Engineers (Finn, 1993) and at the Waterways Experiment Station (Koester, 1992) the "Chinese criteria" were modified by Perlea et al. (1999) for application to index properties as obtained using US standards or similar, as shown in Figure 1. Although

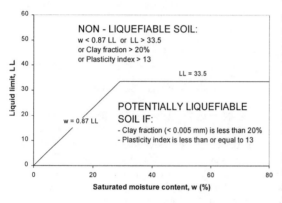

the modifications were relatively minor, at least in one particular case (Sardis Dam, Mississippi) the length of foundation to be consolidated under a dam could be significantly reduced by lowering the limit of problem soil from LL = 35 to LL = 33 (Finn, 1993). Another modification, based on the analysis of Chinese data as reported by Wang (1981), was the addition of the criterion based on plasticity index. Another modification of the original Chinese

Figure 1. Chinese criteria adapted to ASTM definition of soil properties.

criteria should be noted: the significant parameter to be related to LL is the saturated moisture content, not the water content at the time of the investigation, when the potentially liquefiable deposit may not be saturated.

A major deficiency of the "Chinese Criteria" is that the liquefaction susceptibility is not related to the severity of shaking. There is no available information (based on observations in China) on the ground motion characteristics required to trigger this behavior, except that occurrences were reported for strong earthquakes with magnitude 7.6 to 7.8, inducing at the liquefaction sites Modified Mercalli intensities ranging from VII to IX. Based on "Chinese Criteria", any soil meeting them should be considered vulnerable to liquefaction, indifferent of the level of shaking.

An opposite criterion is based on the intensity of shaking, through local (Richter) magnitude (M) and epicentral distance (R), but does not consider the variation in liquefiability due to type and condition of the liquefied deposit; an empirical relationship between M and the maximum epicentral distance of liquefied sites, based on over 50 case histories (Perlea et al., 1999, solid line in Figure 2). On Figure 2 solid circles represent the case histories considered in definition of the envelope; most of them correspond to loose fine sand deposits. Open triangles represent saturated, collapsible loess deposits, that were already on the verge of hydraulic collapse before the earthquake. Open circles were used to plot cases of liquefaction of cohesive soils, other than collapsible loess.

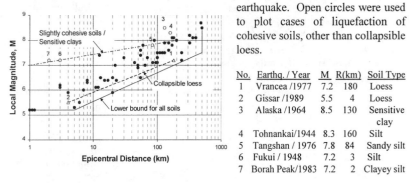

No.	Earthq. / Year	M	R(km)	Soil Type
1	Vrancea /1977	7.2	180	Loess
2	Gissar /1989	5.5	4	Loess
3	Alaska /1964	8.5	130	Sensitive clay
4	Tohnankai/1944	8.3	160	Silt
5	Tangshan / 1976	7.8	84	Sandy silt
6	Fukui / 1948	7.2	3	Silt
7	Borah Peak/1983	7.2	2	Clayey silt

Figure 2. Seismic potential at site vs. liquefaction evidence.

The locations where the events Nos. 1...7 plot suggest a higher located envelope for cohesive soils, except for collapsible loess, as compared with cohesionless soils. It appears, therefore, that triggering liquefaction in cohesive soils requires more released seismic energy than in sands. No liquefaction of cohesive soils (except collapsible loess) was observed as a consequence of earthquake with local (Richter) magnitude less than 7.2. Such seismic events have duration of about 30 seconds or longer; it is believed the cohesive soils need more time than sands for incremental deformations to cumulate until liquefaction state is reached.

Definition of liquefaction

"Liquefaction of soil" is a state of particle suspension resulting from release of contacts between particles. Therefore, the most susceptible to liquefaction are the cohesionless and low plasticity soils, in which the resistance to deformation is mobilized mainly by friction between particles under the influence of confining pressures. The term "liquefaction" has different meanings with respect to various soil conditions. According to Ishihara (1996), the following definitions apply to cohesionless soils:
- For loose sand the (initial) liquefaction is the state of softening in which (indefinitely) large deformation is produced suddenly with (near) complete loss of strength during or immediately following the 100% pore water pressure build-up.
- For medium-dense to dense sand a state of softening (limited liquefaction, cyclic softening, or cyclic mobility) is also produced with the 100% pore water pressure build-up (accompanied by about 5% double-amplitude axial strain) but the deformation does not grow indefinitely large and complete loss of strength does not take place.

- In silty sands or sandy silts the plasticity of fines has a determinant role in liquefiability. Silty soils with non-plastic fines (like many tailings materials) are as easily liquefiable as clean sand. Cohesive fines (as in fluvial deposits) generally increase the cyclic resistance of silty soils. The above definitions of liquefaction for sands are usually applicable to (slightly cohesive) silty soils too.

- Clayey cohesive soils may not loose their strength due to cyclic loading even if they are saturated. By contrary, their undrained strength under dynamic loading is generally higher than static strength. The behavior of clayey materials under cyclic loading is defined by the degradation of strength with the number of cycles and with the corresponding accumulated (residual) strain.

Triggering of liquefaction in cohesive soils

Triggering of liquefaction is generally represented through a series of relationships between cyclic stress ratio (CSR) required to produce 5% double-amplitude axial strain (assumed the onset of liquefaction or cyclic mobility) and the number of cycles (N_1) of an uniform, constant amplitude cyclic loading. $CSR = \tau/\sigma_0'$, where $\pm\tau$ is the maximum cyclically applied shear stress and σ_0' is the effective normal stress acting at the beginning of shaking on the plane where τ is applied. The cyclic (or dynamic) strength is defined as the CSR-value at $N_1 = 10$ (or 20) cycles.

For a given type of soil the relationships $CSR = f(N_1)$ in the series depend on state parameters. The major determinant parameter in sands is the relative density (D_r), but D_r is not an appropriate parameter when the fines content is greater than about 50%. Prakash et al. (1998) found the void ratio (e) a determinant parameter for reconstituted samples of silt, but fabric and plasticity of fines become more important in cohesive soils. The stress history, for example through the overconsolidation ratio (OCR), is a determinant parameter for all soil types.

Cohesive materials, except these with a collapsible structure like highly sensitive clays or collapsible loess, are generally more resistant to liquefaction than cohesionless soils. According to Ishihara (1996) the cyclic strength does not change much for the low plasticity range, below plasticity index, PI = 10, but increase thereafter with increasing PI. Figure 3 demonstrates this trend, although the scatter is important due to variations in type of material, void ratio (0.5 to 1.6), and, possibly, quality of sampling and testing conditions.

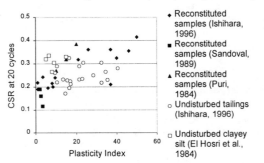

Fig. 3. Dynamic properties of cohesive soils

The criterion of 5% double-amplitude axial strain is convenient to define liquefaction of cohesive soils in triaxial tests, as reliable pore pressure measurements are difficult in cyclic testing of materials with relatively low permeability. However, this selection is also justified by the observation that in cohesive soils cyclically tested the excess pore pressure (Δu) may never reach 100% of the initial confining stress (σ_0'), although the increase in cyclic strain (or accumulated plastic strain) demonstrates significant strength deterioration.

Experimental data in support of this observation were published by El Hosri et al. (1984). Cyclic triaxial tests were performed on undisturbed samples of six different materials, ranging from silty sand to silty clay. Both the number of cycles to 5% axial strain ($N_{l,s}$) and the number of cycles ($N_{l,p}$) necessary for Δu to reach σ_0' were measured. In Figure 4,a the parameter ($N_{l,p}$- $N_{l,s}$)/$N_{l,s}$ has been plotted as a function of plasticity index (PI). It is evident that this parameter generally increases if PI is higher; in the case of the soil with PI = 15, Δu was 85% of σ_0' at the end of the test.

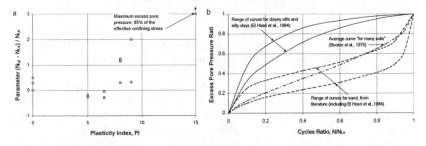

Fig. 4. Rate of pore pressure build-up: a. Number of cycles to $\Delta u = \sigma_0'$ ($N_{l,p}$) compared with number of cycles to initial liquefaction ($N_{l,s}$); b. Difference in characteristic curves for cohesive and cohesionless soils
(data from El Hosri et al., 1984).

The same experimental data (Figure 4,b) lead to another important observation: the excess pore water pressure ratio ($\Delta u /\sigma_0'$) in cohesive soils generally increases much faster than in sand at the beginning of the cyclic loading. However, this increase in pore pressure is not accompanied by same strength degradation as in sand, the cyclic strain increasing significantly close to the end of test only, when the variation in pore pressure is minor.

The strength degradation in soils due to cyclic loading is a complex phenomenon controlled by various factors. In cohesionless soils the most important factor is the pore water pressure build-up, although it is not the only factor. In cohesive soils, the strength against liquefaction does not relay on friction between particles only, but cementation of particles together by electrical and chemical forces plays an important role, which increases with increasing plasticity. Breakage of cementation bonds and re-orientation of particles are significant controlling factors in cohesive soils. Their occurrence is determined by cyclic loading parameters, of which the most important are: number of cycles, their amplitude, and degree of shear stress reversal.

The shear strength degradation with the number of cycles and their amplitude is best represented by the variation of the secant shear modulus ($G = \tau_a / \gamma_a$, where τ_a and γ_a denote the amplitude of shear stress and strain, respectively). In the triaxial test and small deformations, in the elastic range, $\gamma_a = (1+v)\,\varepsilon_a$, where ε_a is the amplitude of axial strain and v the Poisson coefficient equal to 0.5 for saturated soil. In the plastic range $\gamma_a = 1.73\,\varepsilon_a$ (Vucetic and Dobry, 1988). The modulus of cohesive soils tends to decrease significantly with strain amplitude, once a threshold strain is exceeded. This threshold strain is of the order of 5×10^{-5} to 10^{-4}, in contrast to the corresponding behavior in cohesionless soils, in which the modulus reduction starts to occur from a smaller strain, of the order of 10^{-5} (Ishihara, 1996). Vucetic and Dobry (1991) showed that average relations of G/G_0 (where G_0 is the low-amplitude shear modulus) for both cohesionless and cohesive soils demonstrate the importance of the plasticity index on cyclic response (Figure 5).

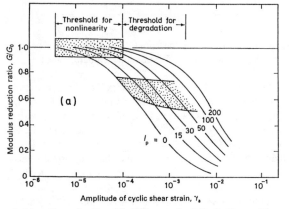

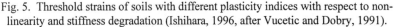

Fig. 5. Threshold strains of soils with different plasticity indices with respect to non-linearity and stiffness degradation (Ishihara, 1996, after Vucetic and Dobry, 1991).

In strain controlled tests (at levels higher than the threshold for degradation) it was observed that secant modulus degradation (both Young's modulus, E, and shear modulus, G) occurred also with the number of cycles. Idriss et al. (1978) define the "degradation parameter", d, as the slope of the straight line representing the variation of E_N / E_1 with the number of cycles, N, in a double logarithmic plot (where E_N and E_1 denote the secant Young's modulus in the first cycle and in the Nth cycle, respectively). Ishihara (1996), referencing studies by Tan and Vucetic (1989) indicated that the degradation parameter can alternatively be defined in terms of the shear modulus: $d = -\log(G_N / G_1) / \log N$ and depends mainly on the strain amplitude, but also on overconsolidation ratio and on plasticity of the cohesive soil. Ishihara (1996) presented the diagram in Figure 6 where two threshold strains are used to characterize soil behavior: the threshold shear strain γ_l differentiating between linearity and nonlinearity and the volumetric threshold shear strain γ_v (defined by Vucetic as the lower limit for the pore water pressure to build-up) separating condition on whether degradation can or cannot occur. The two datum strains have a tendency to increase as the plasticity index of the soil increases.

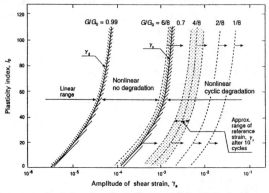

Fig. 6. Summary chart showing the influence of plasticity index on the strain-dependent stiffness degradation (Ishihara, 1996).

According to Ishihara (1996) a relevant index is the reference strain, $\gamma_r = \tau_f / G_0$, that indicates a strain which would be attained at failure stress, if a soil were to behave elastically. Many test results indicated that the reference strain is manifested when the shear modulus is reduced to 40% to 60% of its initial value. Thus, the value of G / G_0 = 4/8 = 0.5 (see Figure 6) may be taken as an average required for the modulus to be decreased to induce the reference strain. It is noted that, in normally encountered state of soils subjected to seismic action, the range of reference strain lies in the domain to the right of the γ_v curve, where cyclic degradation always takes place. In other words, where soils are subjected to cyclic loads with a strain amplitude equal to the reference strain, the soils are already deforming cyclically, accompanied by gradual changes in their properties such as accumulation of volumetric strains or build-up of pore water pressures.

Shear strength of liquefied soil

The post-earthquake behavior of soil and, consequently, the stability of structures founded on liquefied soil, depends on the post-liquefaction (undrained) shear strength of soil. If a conventional shear failure follows cyclic loading, or if liquefaction generates permanent ground deformation and/or flow sliding, depends on the undrained shear strength available at the end of shaking. Depending on the dilatancy properties of the soil, the intensity of shaking, and post-cyclic loading stress path, this shear strength can be higher (dynamic strength is generally higher than the peak undrained strength mobilized by monotonic loading) or lower (critical or residual strength).

Most clayey soils lose a portion of their strength when remolded, due probably to breakage of cementation bonds and reorientation of particles. Clays are considered insensitive when the ratio of undisturbed strength to remolded strength (sensitivity, S) is less than 2, sensitive if S > 4, and moderately sensitive if S is between 2 and 4. A similar drop in strength is experienced by clays subjected to large shearing displacements in drained conditions, from a maximum mobilized "peak strength" to a "residual strength" which is comparable in value with the remolded strength.

In undrained conditions (applicable to short time after cyclic loading) the variation of shear strength with displacement is similar, but affected by the dilatancy properties of soil: during any stage of shearing when the soil has a tendency to dilate, the pore pressure decreases and the shear strength increases; by contrary, a contractive behavior would induce pore pressure increasing accompanied by shear strength lowering. Yamamuro and Lade (1998) applied the principles of steady-state soil mechanics to silty soils. The authors demonstrated that the concept of an unique critical state line, as developed for clean sands, does not apply to silty sands; even a "reverse" behavior was observed by testing silty sand (opposite to that exhibited by clean sand), i.e. increasing dilatancy with increasing confining pressure. This particular behavior was considered due to an unstable structure, with silt grains between larger load-bearing sand grains, which may collapse upon loading, with the silt grains slipping into void spaces. The presence of clay particles in cohesive soils (in particular the relative proportion of round particles and platy clay particles and their mineralogy) controls the dilatancy properties and, therefore, the post-peak behavior. Wroth and Houlsby (1985) explained the departure of the behavior of clay from the concept of critical state soil mechanics by the re-alignment of plate shaped clay particles in a direction parallel to the failure surface; consequently, the mode of deformation changes from "turbulent shear" to "sliding shear". "Residual strength" is associated to the "sliding shear" and only occurs after substantial shear displacements.

Post-cyclic undrained shear strength starts to develop at a point when: (1) excess pore water pressure has built-up and (2) permanent strain has accumulated. With reference to the pore pressure at the end of cyclic loading, Castro and Christian (1976) found that even if the excess pore pressure at the end of cyclic loading was equal to the initial effective confining stress (corresponding to the initial liquefaction or 5% double-amplitude strain) significant undrained strength could be mobilized, due to (negative) pore pressure changes occurring during undrained shear. The strength after cyclic loading was found to depend on the amplitude of cyclic loading: so long as the cyclic strain was less than one-half of the strain required to cause failure in a monotonic test, 90% of the original peak strength was recovered on subsequent monotonic loading. For larger cyclic strains, the strength reduction was more drastic.

Taking into account the above considerations, it is obvious that reliable post-earthquake shear strength of cohesive soils can be determined only by testing good undisturbed samples, modeling the stress path as accurately as possible. Undrained cyclic tests with strain controlled loading, followed immediately by conventional monotonic undrained loading on the same sample, may be convenient. Only if such a test is not an option (because of major difficulties in sampling, handling, and dynamic testing) should the residual strength determined through other methods (e.g. field vane test) be used in design/evaluation of low risk projects; however, it must be recognized that the result of the evaluation may be excessively conservative (or non-conservative, if the pore water pressure plays an important role and the material is highly contractive).

It is noted that the shear strength of the liquefied soil may not be "undrained" at all times during flow failures. Also "post-earthquake" or "post-liquefaction" terms may be misleading, as liquefaction is the process of loss of strength during loading. "Residual strength" without these attributes has a strong connotation to the large strain strength of overconsolidated clays, as first proposed by Skempton. The recommended terminology is "mobilized residual strength" and "steady/critical state strength" to describe field case histories and laboratory test results, respectively (Stark et al., 1998).

Particular liquefiable soils: collapsible loess

Loessial soils are generally wind-deposited sandy/clayey silts in arid areas and have relatively low natural moisture content. They have high porosity (of the order of 45-50%) and most of them are collapsible, i.e. suddenly settle when inundated with water. A typical loess has 50-80% silt size particles, 0-15% clay fraction, liquid limit less than 30, index of plasticity 5 to 15, and the natural moisture content 5-15%. After saturation, the moisture content becomes generally in excess of the liquid limit. A particularity of many loess deposits is the presence of macropores, usually cemented with calcium carbonate, which are assumed to be remains of root holes; they do not increase collapsibility but, by contrary, represent a temporary vertical reinforcement. However, due primarily to the macropores, the vertical permeability (of the order of 10^{-3} to 10^{-4} cm/s) is about 100 times greater than horizontal.

An usual method to determine the collapse potential of soils is the double oedometer test (Figure 7). The difference between the compression curves obtained on a specimen at natural moisture content and on a pre-saturated specimen is an approximation of the settlement by saturation in oedometer, under a given applied pressure, of a specimen at natural moisture content. Loess is considered collapsible when the collapse potential under a load of 300 kPa in oedometer, $i_{m,3} = \Delta e/(1+e_i) > 2\%$ (Stanculescu et al., 1995). However, it was observed that even highly collapsible loess deposits do not settle under the overburden pressure if the layer becoming saturated is less than 6 to 8 meters from surface.

Even with deep loess deposits local saturation may not induce settlement under the overburden pressure in any condition. Due to the significant anisotropy in permeability of loess, the steady state saturation zone under relatively narrow irrigation channels has a limited lateral development (Figure 8). The zone of saturation, although on the verge of collapse, may remain suspended to the massive with low moisture content, through a "silo effect". However, the strength properties of the zone of

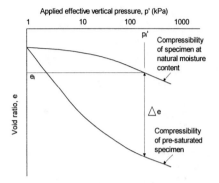

Fig. 7. Double oedometric testing of loess.

saturation are drastically altered (reduced strength, increased compressibility) and the earthquake shaking may provoke the loss of equilibrium. Unlike the loess behavior in oedometer test, collapse does not occur in field until a threshold pressure (corresponding approximately to the own weight of a 6-8 meter layer), is applied. In saturated soil the compressional stress induced by seismic waves is transmitted mostly through the water

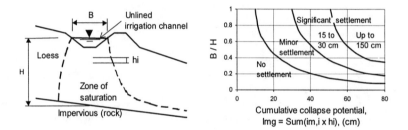

Fig. 8. Qualitative results of observations on irrigation channels and full scale experimental sites on 15-20 meter loess deposits in Romania (after Stanculescu et al., 1995).

in pores, inducing little changes in effective stress. In non-saturated zones, however, vertical acceleration during the earthquake may induce transient pressures in excess of the threshold value on the saturated loess, marginally stable in static condition (e.g. the zone to the right of irrigation channel, in Figure 8), making collapse probable. Through collapse, pore water pressure increases and the low plasticity material liquefies and becomes susceptible to flow. Ishihara et al. (1990) found that the disturbed loess with water content in excess of the liquid limit tends to easily slump, liquefy, and flow.

Particular liquefiable soils: sensitive (quick) clays

Sensitive clay is another example of natural soil deposit already on the verge of collapse before the earthquake that can induce its failure. Sensitivity refers to the loss in undrained shear strength that may develop upon disturbance of the structure of an undisturbed specimen. The measure of sensitivity is the sensitivity ratio, S_t = Peak undisturbed strength / Remolded strength. In laboratory the strength is usually determined through unconfined compression test and in field by vane test. It is usually considered that medium sensitive clays have $S_t = 2$ to 4 and $S_t > 8$ defines quick clay. Because of the flocculated structure and slight cementation, very little strain is needed to break the bonds at particle contacts in sensitive clay. When a saturated sensitive clay is remolded under undrained conditions, the structure is broken down and the pore water increases, with a consequent dramatic reduction in the undrained strength.

There are numerous examples of liquefaction and subsequent mud-flow failure of sensitive clay under non-earthquake actions, especially in overconsolidated deposits. The earthquake, like other sources of disturbance, may initiate slides in gentle sloping deposits of sensitive clay. Among the best documented case histories of seismic induced failures in sensitive clay, were the landslides in Anchorage during the 1964, Alaska earthquake (Stark and Contreras, 1998, Olsen, 1989, Idriss, 1985). The Bootlegger Cove clay, of which failure was responsible for the disaster, had the following average characteristics: LL = 37-39, PI = 14-17, clay size fraction (< 0.002 mm) = 58%, overconsolidation ratio, OCR = 1.2, peak undrained shear strength ratio, $s_u/\sigma_{vc}' = 0.23$ to 0.28, residual undrained shear strength ratio, $s_r/\sigma_{vc}' = 0.06$ to 0.07, and sensitivity ratio, $S_t = s_u/s_r$, of the order of 3 to 4 (as determined with direct simple shear or constant volume ring shear). Through back analyses, it was demon-strated that the mobilized undrained strength in various portions of the landslides varied between at least 80% of s_u and the residual strength, s_r, depending on the earthquake induced permanent deformation (less than 0.15 m and more than 2.5 m, respectively). It is, therefore, impractical to determine triggering criteria, as the earthquake effect depends on many factors defining the pre-earthquake condition of the sensitive deposit.

Youd and Gilstrap (1999) suggest the following simple criteria for sensitive clays susceptible to seismically-induced strength loss: soil types CL or ML, $S_t > 4$, liquidity index > 0.6, moisture content > 0.9 LL, penetration resistance $(N_1)_{60} < 5$ or cone resistance $q_{c1N} < 1$ Mpa, and location in an area where sensitive soil could develop.

Applicability of field tests

The comments that follow are based mainly on the comprehensive investigation of the silty clay and clayey silt alluvial deposits present in the foundation of an embankment dam located in a zone with relatively high potential of seismic activity. Most of the soils in the 20-foot blanket of the dam foundation consist of cohesive soils meeting the "Chinese Criteria", as shown in Figure 9. Although the soil had about 40 years time to consolidate under up to 43 meters of fill, most samples from below the embankment revealed a critical condition with respect to the water content.

Numerous Standard Penetration Test (SPT) were performed at the investigated site, especially in order to characterize the liquefiability of sandy soils existing under the cohesive blanket. The main advantage of SPT in liquefaction evaluation is that both SPT and liquefaction resistance are affected in the same direction (increase or decrease) by variation of relative density, soil fabric, deposit age, K_0, OCR, seismic history.

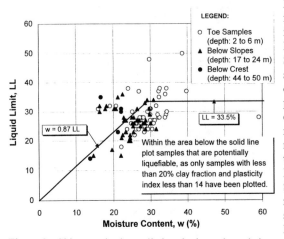

However, the increase in fines content generally increases the resistance to liquefaction but decreases SPT blow count. The same effect, in opposite directions, is due to the plasticity index when in excess of about 10. The most widely accepted method of accounting for the influence of fines on liquefiability is that by Seed et al. (1985). The relationship between stress ratio causing liquefaction and corrected SPT blow count for 35% fines is a conservative

Figure 9. Chinese criteria applied to the investigated site.

value for any cohesionless soil with fines content in excess of 35. It may, however, give either too conservative or non-conservative results when cohesive soils with low plasticity are evaluated. For the toe samples plotted below the solid line in Fig. 9, the average $(N_1)_{60}$ adjusted for fines was about 10, as compared with 13 for data plotted above the line; for samples below the embankment the averages were 24 and 32, respectively, although the difference between moisture contents (which, for saturated materials, are a measure of denseness) and liquid limits were insignificant. This demonstrates that normalization of data as recommended for sand (using the C_N factor) is not applicable to cohesive soils. Fortunately, SPT furnishes also samples and, anytime plasticity is observed, different criteria than Seed-Idriss procedure for sands should be used in evaluation.

There are several empirical methods of liquefiability evaluation using correlations with Cone Penetrometer Test (CPT) results. However, they were developed based on field observations in sands and silty sands, so they should not be applied to cohesive soils, even if the authors extrapolated the contours defining liquefaction susceptibility in the domain of clayey soils. In order to verify the applicability of these available procedures to cohesive soils, electric cone penetration tests and identification tests on samples from adjacent borings have been performed at the investigation site mentioned before. The results of these tests are presented in Fig. 10. It was observed that all materials meeting the "Chinese Criteria" plotted below the line Q_s defined by the equation shown on the graph. All samples that plotted above the Q_s-line were found not meeting the Chinese criteria. The separation by the Q_s-line was not perfect above the Q_m-line, which, according to Robertson (1990) separes sand of silt mixtures; however, materials plotted in this domain qualify for evaluation using criteria for "sands" and "sand mixtures" (e.g. the Seed-Idriss procedure). It results that materials meeting both the two conditions: $Q \le Q_s$ and $Q \le Q_m$ should be considered vulnerable to significant loss of strength due to earthquake shaking, probably meeting the Chinese criteria, and should be tested in laboratory accordingly. Materials with $Q > Q_m$ can be evaluated using criteria for sands and sandy soils. If $Q > Q_s$ and $Q < Q_m$ the material is probably cohesive soil not vulnerable to seismic loading. It should also not be forgotten that the graph in Figure 10 was based on data from one site only and may not be generalized.

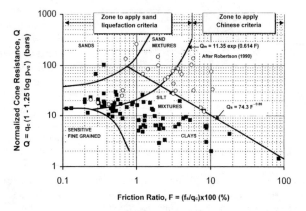

Figure 10. Plot of CPT results in cohesive soil. Solid squares were used to represent samples meeting the Chinese criteria of potentially liquefiable soils and open circles for samples not meeting these criteria.

The attempt to use the available empirical criteria for evaluation of liquefaction susceptibility based on CPT data failed, mainly because friction ratio is a very unstable parameter when the cone resistance varies in a range of low values.

Field Vane Test (FVT) was useful for evaluation of the peak undrained strength (S_u) and the residual undrained strength (S_r). Both parameters were defined as function of the maximum effective overburden pressure experienced by the deposit a time long enough for full consolidation to occur (p_{vo}'). Two types of relationship were determined, as shown in Figure 12,a for comparison with laboratory results: average and conservative (envelope of experimental data). The sensitivity, as defined by the ratio S_u / S_r, was found of the order of 3.

Relevance of laboratory tests

Static (monotonic) undrained triaxial tests were performed on both undisturbed and remolded specimens from the investigation site, to determine the stress-strain behavior and the strain needed for a significant drop in strength. The use of lubricated ends and a height/diameter ratio of about one made possible large deformations of the specimens. The peak undrained strength was measured at axial strains in the range of

15 to 30%, without significant strength decrease thereafter. Unfortunately, 5-inch diameter tube samplers had been used and specimens had to be trimmed, resulting in serious disturbance. More recently (1999) undisturbed sampling and testing of the soft clay deposits meeting the "Chinese criteria" were possible, using adequate equipment, procedures, and skilled operators . Careful measurements of sample dimensions were done during sampling and just before testing to certify insignificant disturbance. The sample and the specimen had the same diameter to avoid necessity of trimming. Variations in volume in excess of about 2% were not tolerated.

Static and dynamic triaxial tests on undisturbed samples obtained using the rational sampling procedures are currently in progress at GEI Consultants laboratory, under the supervision of Dr. Gonzalo Castro. Although the testing program has not been completed yet (as of April 2000) some preliminary findings are already evident:

• Pore pressure increase can correctly be measured if a relatively low frequency of cyclic loading (of the order of 0.25 Hz) is applied and liquefaction "triggering" can be defined as for sandy soils, based on the number of cycles inducing 100% pore pressure ratio or 5% axial strain.

• Samples taken from free field and consolidated under the higher stress condition existing below the embankment developed only about 80% of the resistance to cyclic loading of the samples taken from under the embankment and reconsolidated at the field stress condition (Figure 11,a). This finding emphasizes the significant effect of long term (40 years) consolidation as compared with the consolidation in the laboratory.

• The increase in pore pressure ratio versus the number of cycles normalized by the number of cycles to 5% strain was faster at the beginning of cycling than generally for sands, as presented in Figure 11,b (curves for 12 tests and their average are shown).

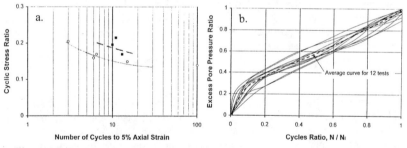

Figure 11. Results of cyclic triaxial tests: a. Liquefaction triggering [the average consolidation stress, $p' = (\sigma_1 + \sigma_3)/2$, was used in cyclic stress normalization]; open circles represent samples taken from free field and solid squares samples from under the embankment. b. Pore pressure ratio evolution during the test.

• Most of cyclic testing were performed after anisotropic consolidation and were followed by monotonic shearing in compression, after at least 5% strain occurrence in the dynamic phase and without dissipating the pore pressures induced by cyclic loading. Figure 12 presents the results of the monotonic undrained compression phase.

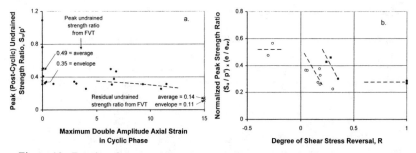

Figure 12. Results of post-cyclic monotonic compression tests: a. Peak undrained strength ratio versus the maximum double amplitude strain reached at the end of cyclic phase (static tests were plotted at zero strain; results of FVT were also plotted for comparison). b. Peak strength ratio normalized for an average void ratio of 0.75 versus the degree of shear stress reversal, R; open circles represent samples taken from free field and solid squares samples from under the embankment.

• The peak strength ratio for strains less than about 7% were at the level measured in field vane tests. At approximately 10% maximum cyclic strain, some loss of undrained strength was observed. However, even after having been subjected to 10% maximum cyclic strain plus an additional 20% under monotonic loading, the soil still retained a peak strength more than double the residual strength measured in field vane tests.

• The decrease in peak undrained shear strength appears to be also a function of the degree of shear stress reversal, R. The parameter R is defined as $R = (\Delta\tau - \tau_s) / (\Delta\tau + \tau_s)$ where $\Delta\tau$ and τ_s are the applied cyclic shear stress and the initial shear stress on a 45° plane, respectively. $R = 1$ for symmetric reversal (isotropic consolidation) and R is negative when there is no shear stress reversal. Although some other parameters affected the results, the post-cyclic shear strength was higher when the cyclic phase was conducted without shear stress reversal.

Conclusions

In general, cohesive soils appear to be more resistant to liquefaction than fine sands, when their states of denseness are similar. The strength degradation is dependent on cyclic strain magnitude or total accumulated displacement. As accumulation of displacement requires time, it is apparent that seismic action of relatively long duration is required for significant strength loss of cohesive soils. However, the residual strength mobilized under large deformations may be smaller than that of sandy soils with similar state of denseness.

The post-liquefaction behavior of cohesive soils depends primarily on their dilatancy properties and the characteristics of the cyclic loading: number of cycles, amplitude, frequency, and magnitude of shear stress reversal. Taking into account the multitude of factors affecting liquefiability and post-cyclic undrained strength of cohesive soils, the best way of evaluation of their behavior under cyclic loading is by test.

Major difficulties in accurately replicate the fabric of field conditions through reconstituted laboratory specimens, make testing of undisturbed samples a necessity. Variations in quality of undisturbed samples and testing methodologies make unreliable attempts to correlate data from various sources.

The recommended test is cyclic loading in undrained conditions followed immediately (without drainage permitted) by monotonic loading. Pre-existence of shear stresses on the planes of maximum applied cyclic shear stress should be given consideration, as shear stress reversal is an important factor in liquefaction triggering and evolution of strength degradation with accumulated displacement. Field tests (vane, cone penetration, shear wave velocity measurements) may be useful in conservative evaluation of stability of non-critical structures on liquefiable soil deposits.

References

Booker, J.R., Rahman, M.S., and Seed, H.B. (1976). "GADFLEA, a computer program for the analysis of pore pressure generation and dissipation during cyclic or earthquake loading." *Report No. EERC 76-24,* University of California, Berkeley, California.

Castro, G. and Christian, J.T. (1976). "Shear strength of soils and cyclic loading." *Journ. of Geotechnical Engrg.* ASCE, Vol. 102, GT9, 887-894.

El Hosri, M.S., Biarez, J., Hicher, P.Y. (1984). "Liquefaction characteristics of silty clay." *Proc., Eighth World Conf. on Earthquake Engrg.,* San Francisco, California, Vol.3, 277-284.

Finn, Liam W.D. (1993). "Evaluation of liquefaction potential." *Soil Dynamics and Geotechnical Earthquake Engineering,* Seco e Pinto (ed.), Balkema, 127-157.

Hall, J.W. (Upton Close) and McCormick, E. (1922). "Where the mountains walked." *The National Geographic Magazine,* Vol. XLI, No.5, 445-464.

Hynes, M.E. (1999). "Dam lessons learned from major earthquakes this year." *Private Communication.*

Idriss, I.M. (1985). "Evaluating seismic risk in engineering practice." *Proc., 11th Int. Conf. on Soil Mechanics and Foundation Engrg.* San Francisco, Vol.1, 255-320.

Idriss, I.M., Dobry, R., and Singh, R.D. (1978). "Nonlinear behavior of soft clays during cyclic loading." *Journ. of Geotechnical Engrg.,* ASCE, Vol. 104, GT12, 1427-1447.

Ishihara, K. (1985). "Stability of natural deposits during earthquakes." *Proc., 11th Int. Conf. on Soil Mechanics and Foundation Engrg.* San Francisco, Vol.1, 321-376.

Ishihara, K. (1984). "Post-earthquake failure of a tailings dam due to liquefaction of the pond deposit." *Proc., Int. Conf. on Case Histories in Geotechnical Engrg.,* St. Louis, Missouri, Vol.3, 1129-1143.

Ishihara, K. (1996). *"Soil behaviour in earthquake engineering",* Clarendon Press, Oxford, Great Britain, 350 pages.

Ishihara, K., Okusa, S., Oyagi, N., and Ischuk, A. (1990). "Liquefaction induced flow slide in the collapsible loess deposit in Soviet Tajik. *Soils and Foundations,* Vol. 30, No.4, 73-89.

Japan Society of Civil Engineers (1977). *"Earthquake resistant design for civil engineering structures. Earth structures and foundations in Japan."*

Kishida, H. (1969). "Characteristics of liquefied sands during Mino-Owari, Tohnankai and Fukui earthquakes." *Soils and Foundations,* Vol.9, No.1, 75-92.

Koester, J.P. (1992). "The influence of test procedure on correlation of Atterberg limits with liquefaction in fine-grained soils." *Geotechnical Testing Journal,* 15(4): 352-361.

Miura, S., Kawamura S., and Yagi, K. (1995). "Liquefaction damage of sandy and volcanic grounds in the 1993 Hokkaido Nansei-Oki earthquake." *Proc. 3rd Int. Conf. on Recent Advances in Geotechnical Earthq. Engrg. and Soil Dynamics,* St. Louis, Missouri, Vol.1, 193-196.

Obermeier, S.F., Gohn, G.S., Weems, R.E., Gelinas, R.L., and Rubin, M. (1985). "Geologic evidence for recurrent moderate to large earthquakes near Charleston, South Carolina." *Science,* 1(227), 408-411.

Olsen, H.W. (1989). "Sensitive strata in Bootlegger Cove formation." *Journ. of Geotechnical Engrg.,* ASCE, 115(9), 1239-1251.

Perlea, V.G., Koester, J.P., and Prakash, S. (1999). "How liquefiable are cohesive soils?" *Proc. Second Int. Conf. on Earthq. Geotechnical Engrg.,* Lisbon, Portugal, Vol.2, 611-618.

Prakash, S., Guo, T., and Kumar, S. (1998). "Liquefaction of silts and silt-clay mixtures." *Proc. 1998 Spec. Conf. on Geotech. Earthq. Engrg. and Soil Dynamics,* Seattle, WA, 1:327-348.

Puri, V.K. (1984). *"Liquefaction behavior and dynamic properties of loessial (silty) soils."* Ph.D. Thesis, University of Missouri-Rolla, Missouri.

Robertson, P.K. (1990). "Soil classification using the cone penetration test." *Canadian Geotechnical Journal,* 27(1), 151-158.

Sandoval, J. (1989). *"Liquefaction and settlement characteristics of silt soils."* Ph.D. Thesis, University of Missouri-Rolla, Missouri.

Seed, H.B., Idriss, I.M., and Arango, I., (1983). "Evaluation of liquefaction potential using field performance data." *Journ. of Geotechnical Engrg.,* ASCE, 109(3), 458-482.

Seed, H.B., Tokimatsu, H., Harder, L.F., and Chung, R.M., (1985). "Influence of SPT procedures in soil liquefaction resistance evaluations." *Journ. of Geotechnical Engg, ASCE,* 111(12), 1425-1445.

Seed, H.B. and Wilson, S.D. (1967). "The Turnagain Heights landslide, Anchorage, Alaska." *Journ. Soil Mechanics and Foundation Div.,* ASCE, 93(SM4), 325-353.

Seed, H.B., Seed, R.B., Harder, L.F., and Jong, H.L. (1989). "Re-evaluation of the Lower San Fernando dam - Report 2: Examination of the post-earthquake slide of February 9, 1971." *Contract Report GL-89-2, U.S. Army Engineer WES, Vicksburg, Mississippi.*

Stanculescu, I., Bally, R.J., Antonescu, I.P., and others (1995). "Some civil engineering aspects concerning loessial collapsible soils in Romania." *Geotechnical Engineering in Romania,* Technical University of Civil Engineering, Bucharest, Romania, 21-42.

Stark, T.D. and Contreras, I.A. (1998). "Fourth Avenue landslide during 1964 Alaskan earthquake." *J. Geotech. and Geoenvironmental Engrg.,* ASCE, 124(2), 99-109.

Stark, T.D., Olson, S.M., Kramer, S.L., and Youd, T.L., editors, (1998). "Working group discussions" and "Summary and recommendations", *Proc., "Shear Strength of Liquefied Soils" Workshop,* Urbana, Illinois, April 1997, National Science Foundation, 58-80.

Tan, K. and Vucetic, M. (1989). "Behaviour of medium and low plasticity clays under simple shear conditions." *Proc. Fourth Int. Conf. on Soil Dynamics and Earthq. Engrg.,* Mexico City, 131-141.

Tohno, I. and Yasuda, S. (1981). "Liquefaction of the ground during the 1978 Miyagiken-Oki earthquake." *Soils and Foundations,* 21(3), 18-34.

Tuttle, M.P. and Seeber, L. (1989). "Earthquake-induced liquefaction in the northeastern United States: historical effects and geological constraints." *Annals of the NY Academy of Sciences,* 558, 196-207.

Vucetic, M. and Dobry, R. (1988). "Degradation of marine clays under cyclic loading." *Journ. of Geotechnical Engrg.,* ASCE, Vol. 114, GT2, 133-149.

Vucetic, M. and Dobry, R. (1991). "Effect of soil plasticity on cyclic response." *Journ. of Geotechnical Engrg.,* ASCE, Vol. 117, No. 1, 89-107.

Wang, W. (1979). "Some findings in soil liquefaction." *Report Water Conservancy and Hydro-electric Power Scientific Research Institute,* Beijing, China, 1-17.

Wang, W. (1981). "Foundation problems in aseismatic design of hydraulic structures." *Proc., Joint U.S.-P.R.C. Microzonation Workshop,* Harbin, China, 15-1 - 15-13.

Wesnousky, S.G., Schweig, E.S., and Pezzopane, S.K. (1989). "Extent and character of soil liquefaction during the 1811-1812-New Madrid earthquakes." *Ann. of the NY Academy of Sciences,* 558: 208-216.

Wroth, C.P. and Houlsby, G.T. (1985). "Soil mechanics - Property characterization and analysis procedures." *Proc., Eleventh Int. Conf. on soil Mechanics and Foundation Engrg.,* San Francisco, California, Vol. 1, 1-55.

Yamamuro, J.A. and Lade P.V. (1998). "Steady-state concepts and static liquefaction of silty sands." *Journ of Geotechnical and Geoenvironmental Engrg.,* ASCE, Vol. 124, No. 9, 868-877.

Youd, T.L. and Gilstrap, S.D. (1999). "Liquefaction and deformation of silty and fine-grained soils." *Proc. Second Int. Conf. on Earthq. Geotechnical Engrg.,* Lisbon, Portugal, Vol.3, 1013-1020.

Youd, T.L., Harp, E.L., Keefer, D.K., and Wilson, R.C. (1985). "The Borah Peak, Idaho earthquake of October 28, 1983 - - liquefaction." *Earthquake Spectra,* Earthquake Engrg. Res. Inst., 2(1): 71-89.

Effect of Non-Plastic Fines on Undrained Cyclic Strength of Silty Sands

S. Thevanayagam[1], M. Fiorillo[2] and J. Liang[2]

Abstract

Whether the presence of silt adversely or beneficially affects liquefaction and the collapse potential of silty soils and how to evaluate cyclic strength behavior of a sand containing different silt contents are contentious issues. The purpose of this work is to investigate this question. Stress controlled undrained cyclic triaxial tests were conducted on specimen prepared by mixing a sand with silt in different proportions. The cyclic stress ratio (CSR=0.2) and confining pressure (100 kPa) were maintained constant. Relationship between no. of cycles required to cause liquefaction (at 5% axial strain) versus void ratio, and newly introduced equivalent void ratio indices based on intergrain contact density considerations are presented. Cyclic strength correlates well with the latter indices.

Introduction

Recent earthquake case histories indicate that natural soils and man-made sandy deposits that contain a significant amount of finer-grains (silty sands, clayey sands) and/or gravel do liquefy and cause lateral spreads (Seed et al. 1983, Seed and Harder 1990, JGS 1996). Experience gained from past studies on clean sands based on void ratio or relative density as index parameters does not always directly translate to such broadly graded soils. Recognition of this has lead to several laboratory and field studies to evaluate the effects of increasing silt or gravel content on: (a) cyclic strength, (b) collapse potential, (c) steady state strength, (d) shear wave velocity, etc. Results from laboratory studies on clean sands mixed with non-plastic silts or plastic fines show that, at the same (global) void ratio, the steady state strength and cyclic strength of silty sand decreases with an increase in fines content (Chang 1990, Chameau and Sutterer 1994, Georgiannou et al. 1990,91a-b, Vaid 1994, Koester 1994, Pitman et al. 1994, Singh 1994, Zlatovic and Ishihara, 1997, Thevanayagam et al. 1996, Yamamuro and Lade 1998). An example is shown in Fig.1a for a soil

---

[1] *Assoc. Prof., Dept. of Civil, Struct., & Env. Eng., State Univ. of New York, 212 Ketter, Buffalo, NY 14260; Email: theva@eng.buffalo.edu*
[2] *Graduate student, Dept. of Civil, Struct., & Env. Eng., State Univ. of New York*

prepared by mixing a sand [M = Medium Sand] with a silt (PI=4) at nearly constant global void ratio (e=0.558) and confining stress (104 kPa) but at different silts content (0 to 60% by weight). Fig.1b shows the number of cycles to reach initial liquefaction versus silts content for the same sand-silt mix, along with the data for two other sands [F = Fine Sand, W = Well Graded Sand] mixed with the same silt. [Fig.1b can also be expressed in the form of Fig. 7d, which will be discussed later]. As observed in these figures, beyond a certain transition range the trend in decrease of strength reverses and the strength increases with further increase in silts content. The transition fines content range is about 20 to 30% for non-plastic fines (Vaid 1994, Kuerbis et al. 1988, Singh 1994, Koester 1994). It is less than 20% for clayey (plastic) fines (Georgiannou et al. 1991a-b). The physical meaning of the transition fines content is not clear. The conclusions in the literature on whether the presence of fines is beneficial or not is contentious. Similar concerns prevail regarding gravely soils (Evans and Zhou 1995).

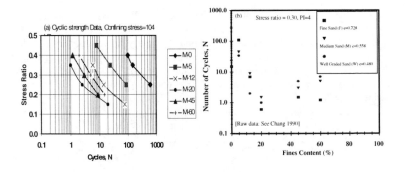

Fig.1. Influence of finer grain content on cyclic strength – (a) Medium Sand, (b) Three Sands

In this regard, recently, it has been noted that physical nature of silty sands and gravely sands is entirely different from clean sand (Thevanayagam 1998a-b, 1999, Thevanayagam and Mohan 1998). As the void ratio and proportion of the coarser and finer grains content of these soils change the nature of their microstructure also changes. The relative participation of the particles of very different sizes in the *internal interparticle contact force chain* also changes. Due to particle size disparity and availability of pores larger than some particles, at low finer grains content some of the finer particles may remain inactive or move between pores without significantly affecting or contributing to the force chain. Yet they contribute to the global void ratio. Alternately when there are sufficient amount of finer grains the coarser grains become dispersed contributing much less to the force chain than to the global void ratio. Global void ratio e ceases to be an index to represent the nature of contact density of active particles. The traditional use of e to compare the behavior of soils containing different amounts of fines content ceases to be valid. The same holds for relative density.

In general the stress-strain behavior, liquefaction potential, and fragility of granular mixes are affected by a critical combination of *intergranular and interfine contacts* and the physical and physico-chemical interactions thereof. The combined effects of intergranular and interfine contacts must be delineated in dealing with silty sands and gravely soils in understanding the mechanisms leading to liquefaction and post-liquefaction deformation, and the mechanical response of the media in general. New indices of active contacts are needed to represent the nature of intergrain contacts in order to characterize the behavior of such soils. It is thought that recognition of these factors may help to bring about a rational method for liquefaction potential assessment of silty and gravely soils.

Using a two-sized particle mix as a model, this paper highlights the nature of the microstructure of granular mixes. Based on this such granular mixes are classified into certain groups (Fig.2) depending on the relative frictional contributions at the intergranular and interfine grain contact level. The range of void ratio and fines content where each group (Fig.2) belongs to is conveniently shown in a global void ratio versus fines content diagram (Fig.4). Taking into account possible interactions between coarser and finer grains, two equivalent intergranular $(e_c)_{eq}$ and interfine $(e_f)_{eq}$ (Thevanayagam 1998b) void ratios (Fig.3) are introduced as primary indices of contact density for granular mixes at low and high finer grain contents. The usefulness of these indices to characterize the observed cyclic strength behavior of granular mixes is evaluated.

Conceptual Framework

Microstructure

Consider a two-sized granular mix. The microstructure of this granular mix can be constituted by infinite different ways. Each one of them leads to a different internal force chain network among particles and hence each exhibits a different stress-strain response during shear. Among infinite variations, a few extreme limiting categories of microstructure and the relevant roles of coarser and finer grains are as follows.

Case-i: The first category (Fig.2a) is when the finer grains are fully confined within the void spaces between the coarser-grains with no contribution whatsoever in supporting the coarser grain skeleton. Finer grains are inactive (or secondary) in the transfer of inter particle forces. They may largely play the role of "filler" of intergranular voids. The mechanical behavior is affected primarily by the coarser grain contacts. During deformation the finer grains may move from one pore space to another without significantly contributing to the mechanical response of the soil. This requires that the finer grain particle size (d) is much smaller than the pore size between the coarser grains and that the intergranular pore space is not completely filled with the finer grains. Typically this requires that the coarser grain size (D) is at least 6.5 times larger than the finer grain size, and that the finer grain content (FC) is less than a certain threshold value (FC_{th}). This category is called *case-i*. Even at low

FC, if the size disparity R_d (=D/d) is not very large, the finer grains cannot freely move through the inter-coarser granular voids; They also tend to participate in the force chain and actively contribute to the stress-strain response.

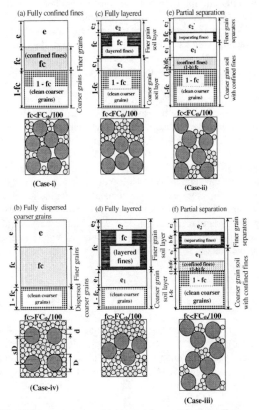

Fig.2 Microstructure and intergranular matrix phase diagram

Cases-ii and iii: Consider changing the microstructure shown in Fig.2a in two ways: (1) alter the position of some of the finer grains, or (2) add more finer grains. The consequences are significant. If one alters the position of some of the finer grains, while maintaining the finer grain content the same, the microstructure corresponding to the second and third categories shown in Figs.2e and f are obtained with a concurrent increase in global void ratio. Essentially, the microstructure in Figs.2e-f is made up of partial layering and partial separation of coarser grains by the finer grains along with confined finer grains within the voids between the coarser grains. Some of the finer grains become active participants in the internal force chain. These finer grains are termed the *'separating fines'* in Figs.2e and f. In Fig.2e, the finer grains may be supporting the coarser-grain skeleton that is otherwise unstable.

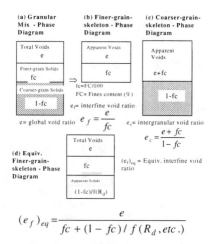

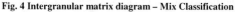

Fig.3 Intergranular contact indices

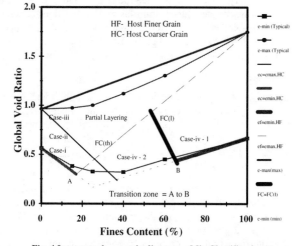

Fig. 4 Intergranular matrix diagram – Mix Classification

They act as a load transfer vehicle between "some" of the coarse-grain particles in the soil-matrix while the remainder of the fines play the role of "filler" of voids. They may dominate the initial stress-strain behavior depending on the type of finer grains (plastic or non-plastic). In Fig.2f, the finer grains may play an active role of *"separator"* between a significant number of coarse-grain contacts and therefore begin to dominate the strength characteristics. Coarser grain skeleton is virtually unstable without the finer grains. These two categories of microstructure are called cases ii and iii, respectively. Case-ii is a transition between cases i and iii. Theoretically case-iii occurs at an intergranular void ratio exceeding the maximum void ratio ($e_{max,HC}$) achievable for the 'pure' coarser grain soil.

Cases-iv-1 and iv-2: On the other hand if one increases the finer grains content sufficiently, one gets the fourth category (Fig.2b). It occurs naturally when sufficient finer grains are present making active contacts among themselves. The coarser grains begin to disperse in the finer grain matrix. Transition from Fig.2a to Fig.2b occurs when the finer grains content (FC) exceeds beyond the threshold fines content (FC_{th}). When $FC > FC_{th}$ the finer grains begin to play a rather important role while the role of coarser grains begin to diminish. The fines may carry the contact and shear forces while the coarser grains may act as *reinforcing elements* embedded within the finer grain matrix. The effect of coarser grains cannot be completely neglected until they are separated sufficiently apart. This imposes a limiting fines content FC_l. There exists a transition zone between FC_{th} and FC_l before the behavior of the soil mix is entirely governed by the finer grains. This is called case-iv-2 whereas the case corresponding to $FC > FC_l$ is called case-iv-1. The size disparity constraint discussed before for cases i to iii needs not be satisfied for case-iv.

The fifth category (Figs.2c-d) is when the coarser and finer grains constitute a fully layered system where the coarser grain layers have no fines contained in them and vice versa. This is called case-v. It is also possible to create a composite system that contains some of the cases i through v. The figures 2a, c, e and f are more relevant at low finer grains content. Figs. 2b and d are relevant at high finer grains content.

Conceptually Fig.4 shows the regions belonging to the four cases i through iv confined by various transition boundaries. The transition lines corresponding to the threshold FC_{th} and limiting FC_l fines contents may be estimated using:

$$FC_{th} \leq \frac{100 e_c}{1 + e_c + e_{max.HF}} \% = \frac{100 e}{e_{max.HF}} \% \; ; \qquad FC_l \geq 100 \left[1 - \frac{\pi(1+e)}{6s^3}\right] \% = 100 \left[\frac{\frac{6s^3}{\pi} - 1}{\frac{6s^3}{\pi} + e_f}\right] \% \geq FC_{th} \,; e_f \leq e_{max.HF} \quad (1)$$

where $e_c = (e + fc)/(1 - fc)$, $e_f = e/fc$, $fc = FC/100$, $s = 1 + a(d/D) = 1 + a/R_d$, and $a = 10$ (approximately). The rationale behind the equation for FC_{th} is that once the interfiner grain void ratio e_f drops below $e_{max,HF}$ (the maximum void ratio achievable for the 'pure' finer grain soil) the finer grains begin to make active contacts among themselves and contribute to the force chain. The reasons leading to the derivation of the expression for FC_l may be attributed to the observations of Roscoe (1970) that the zone of influence of shear is about 10 times the diameter of particles. The various other boundaries refer to the maximum and minimum void ratio profiles: $e = e_{max,HC} + (e_{max,HF} - e_{max,HC}) fc$; $e_c = e_{max,HC}$; $e_f = e_{max,HF}$; $e_c = e_{min,HC}$; and $e_f = e_{min,HF}$.

Contact Indices and Mechanical Behavior

Recently the nature of intergrain contacts within such granular mixes has been studied in some detail. Results show that up to $FC=FC_{th}$ the finer grains can, but not necessarily, remain within the intergranular voids. Primarily the intergranular contacts between the coarser grains affect the mechanical behavior with secondary effects by the finer grains. Hence, neglecting the effects of fines, the *inter-coarser grain void ratio* e_c (Fig.3c) may be used as an index of active contacts. The magnitude of e_f may be used as an index to assess the secondary effects by the finer grains. If the secondary effects are included, the relevant *equivalent intergranular contact index void ratio* would be of the form $(e_c)_{eq}$ $(=[e+(1-b)fc]/[1-(1-b)fc])$ (Fig.2). At very high FC ($>FC_{th}$), neglecting the effects of dispersed coarser grains, the *interfine void ratio* e_f ($=e/fc$, Fig.3b) may be used as an index of active contacts. If the secondary effect of the coarser grains are included, recent work shows that the relevant *equivalent interfine contact index void ratio* is of the form $(e_f)_{eq}$ $=e/[fc+(1-fc)/(R_d)^m]$, Figs.2b and 3d, $0<m<1$, $e_c>e_{max,HC}$ (Thevanayagam 1998b, Thevanayagam 2000).

The aforementioned contact indices can be used as aids to predict the trends of the stress-strain characteristics, liquefaction potential, and fragility of silty or gravely soils (prepared by the same method at the same confining stress) *relative* to that of the host coarser grain soil or the finer grain soil. Fig.5 shows a schematic diagram for *hypothetical* specimens satisfying the following specific constraints: (1) an increase in global void ratio e at the same fines content [specimens 1,2,3], (2) an increase in fines content at a constant global void ratio e [4,2,5,6,7,13], (3) an increase in fines content at the same intergranular void ratio e_c [8,1,9 or 3,10 or 14,2,15; $FC<FC_{th}$], or (4) an increase in coarser grain content at the same interfine void ratio e_f [11,7,12; $FC>FC_{th}$].

The anticipated trends in number of cycles (N) required to cause initial liquefaction at the same cyclic stress ratio are *schematically* shown in Fig.5 [N versus FC]. In (1) both intergranular and interfine void ratios increase with concurrent reduction in inter-coarser grain contacts. Therefore the soil becomes weaker. In (2) while e_c increases e_f decreases. Viz. the inter-coarser granular contacts decrease while the interfine contacts increase. Hence initially the soil is expected to weaken [4,2,5] followed by a transition in the vicinity of $e_f=e_{max,HF}$ ($FC=FC_{th}$). The soil becomes stronger beyond that [6,7,13]. In (3), the increase in cushioning effect by the fines is manifested leading to a slight increase in strength [3,10, case-iii]. The specimen 10 is expected to be somewhat more resistant to collapse than the specimen 3. This effect, however, diminishes gradually if the soil becomes denser in terms of e_c [case-ii]. The reason is that when e_c is small (dense coarser grains) the relative effect of fines is less appreciable compared to the direct coarser-coarser grain contact resistance until e_f becomes sufficiently low. In (4) when the soil is at fines content less than FC_l but greater than FC_{th} [11,7,12] the reinforcement effect of the coarser grain 'inclusions' may affect the stress strain behavior. The specimen 11 is expected to be stronger than 7. Again this reinforcement effect may become relatively small compared to the direct finer-grain-to-finer-grain contact resistance when e_f is small (dense interfine contacts). Once FC exceeds FC_l the reinforcement effect is expected to be small.

Primarily the interfine contacts and e_f are expected to affect the soil behavior [7,12]. Without elaboration, the remaining figures show the relative trends for N for various cases i through iv plotted against the relevant contact indices (e_c, e_f, and $(e_f)_{eq}$.

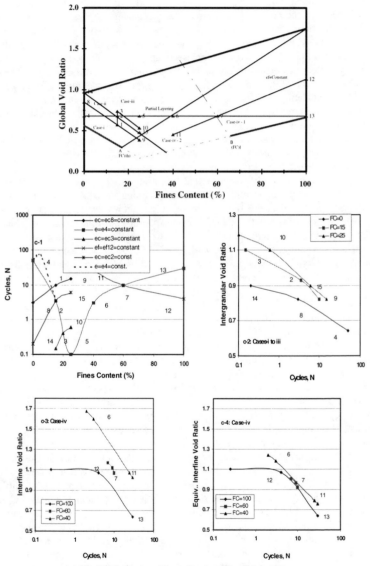

Fig.5 Relative trend in cyclic strength – Schematic.

Experimental Evaluation

This section presents an experimental evaluation of the above conceptual framework. For this purpose the experimental data reported by Chang (1990), shown before in Figs.1a-b, and additional new experimental data generated as a part of this study are used. The latter experiments involve undrained triaxial cyclic tests on large size specimens (typically 74 mm diameter and 150 ~ 160 mm height) prepared using a single host sand (OS55) (OS-F55, Foundry Sand from US Silica Company, Illinois, e_{max}=0.80, e_{min}=0.60, d_{50}=0.24mm) mixed with different amounts of non-plastic crushed silica fines (Sil co sil #40, 99.9% passing sieve #200, d_{50}=0.007mm) at (a) 0%, (b) 15%, (c) 25%, and (d) 60% fines by dry weight.

Cylindrical specimens were prepared by placing soils in four layers or eight layers in a triaxial mold using dry air deposition method or moist tamping method. The mold was filled with the soil by layers and compacted by gentle tamping until reaching a specified target void ratio. The procedure was similar for each layer until a specified target void ratio was reached. Then the specimens were percolated with carbon dioxide for about 5 minutes. After that, de-aired water was allowed to flow from the bottom of the specimen towards the top if the specimen was prepared by dry deposition method (If the specimen was prepared by moist tamping method, this step was omitted). Then the specimens were back-pressure saturated. The net volume of the water introduced into the specimen was measured accurately. Finally the specimen was set up on the triaxial test apparatus to continue the saturation until the B-value ($=\Delta u/\Delta\sigma_c$) was typically greater than 0.95. Following this, the specimens were isotropic consolidated to a constant effective consolidation stress (σ'_c=100 kPa. Cyclic stress ratio ([CSR=$\pm\Delta\sigma_1/(2\sigma'_c)$]=0.2) controlled undrained triaxial cyclic tests were done at a frequency of 0.2 Hz or 1 Hz. The pore pressure, axial load, and axial deformation were recorded using a built-in data acquisition system. The final void ratio of each specimen was calculated based on the weight of the dry solid grains in the specimen, the net volume of water introduced into the specimen during saturation, and the measured volume change data during consolidation.

Reinterpretation of Existing Data

Fig.6 shows the intergranular matrix diagram for one of the three host sand/silt mixes [(1) Fine (F), (2) Medium (M), and (3) Well graded (W)] tested by Chang (1990), reported in Fig.1a. Each sand mix was tested at 0, 5, 12, 20, 45, and 60% fines content. The specimens were prepared by moist tamping method. All specimens were consolidated to the same initial confining stress (104kPa). The specimens for each soil mix were tested at nearly constant global void ratio: e=0.728 for F, e=0.558 for M, and e=0.480 for W, respectively. These void ratios correspond to about 50% relative density of the respective parent sands. No data were available for $e_{max,HF}$ and $e_{min,HF}$ for the silt. For qualitative discussion purposes the threshold boundary may be estimated assuming typical values of $e_{max,HF}$ =1.5 and $e_{min,HF}$ =0.6. Also for illustration purposes the R_d (defined as D_{50}/d_{50}) values for the soils F, M, and W were estimated to be about 15, 40, and 40 based on a typical value for d_{50} of silts.

An examination of Fig.6 and calculated index void ratios e_c, e_f, would readily reveal the cases each specimen belongs to and the expected behavior of each granular mix relative to one another. For all three mixes the specimens at FC ranging from 0 to 20% fall within the category of cases i to iii. Cyclic strength (N) correlates with intercoarser grain contact index e_c for these specimens (Fig.7a). The three different trends observed in this figure is due to differences in parent (host) sand type (F,M,W). Due to lack of sufficient data no comparison is made against $(e_c)_{eq}$. The specimens at 40 and 60% fall in the category of case-iv. N correlates with interfine contact index e_f or $(e_f)_{eq}$ for these specimens (Figs.7b-c). Cyclic strength for these soils is significantly dependent on e_f. The data for each fines content (45% and 60% separately) correlates well with e_f and fall in a (separate for 45% and 60%) narrow band regardless of the parent sand type. The separate narrow bands for the soils mixed with 45% fines and 60% fines are due to the differences in the degree of reinforcement effect. At the same e_f, the cyclic strength is higher for the soil with higher sands content. The reinforcement effect is higher at 45% fines content than at 60% which is approaching the limiting fines content FC_l. Nevertheless, at low e_f (dense interfine contacts) the relative effect of reinforcement becomes less appreciable compared to direct interfine contact friction. The separate bands for 45 and 60% merge together at low e_f. Fig.7c shows the same data, plotted against equivalent interfine void ratio $(e_f)_{eq}$ [assuming m=0.45]. Interestingly the number of cycles required to cause liquefaction correlates with $(e_f)_{eq}$ for all cases of iv. The reason for this is that all sands were mixed with the *same silt*. Hence, once FC exceeds the threshold value, all soil mixes are affected by the same silt except for the minor differences due to the presence of different size coarser grains (F,M,W). Hence a single narrow band is obtained in Fig.7c. It would be more revealing if (if available) the data for the silt is superimposed in this figure permitting a comparison of the kind shown in Fig.5 [$(e_f)_{eq}$ versus N].

Fig.7d shows the same cyclic strength data shown in Fig.1b. The e_c and e_f values for each specimen and the respective cases each specimen belongs to are also shown in this figure. As the fines content increases, at the same global void ratio e, the e_c increases and e_f decreases. Initially the e_f remains high to be of any significance, alone. With increase in FC the soil mixes move gradually from Case-i to Case-ii to Case-iii and then cross over to Case-iv-2. So is the behavior of the soils. Initially the strength decreases due to reduced intergranular contacts with increase in e_c with little or secondary contribution from the fines. As the soil moves beyond the threshold transition zone (FC_{th} at $e_f = e_{max,HF}$) and enters the zone for Case-iv-2 the influence of e_f becomes important with some reinforcement effect by the coarser grains. The $(e_f)_{eq}$ becomes the primary contact index void ratio. The trend reverses and the strength begins to increase with further increase in fines content. The transition fines content FC_{th} is slightly different for each soil mix. Theoretically it corresponds to the intersection of the constant global void ratio line with the $e_{max,HF}$ line. For the same host fines, typically a soil mix at a smaller global void ratio will cross the $e_{max,HF}$ line and reach Case-iv-2 at a smaller fines content than a soil mix at a higher global void ratio. Hence, the soil W (at e=0.480) reaches this transition at smaller fines content than the soil F (at e=0.728).

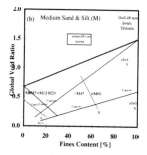

Fig.6 Intergranular matrix diagrams: Medium Sand-Silt Mix

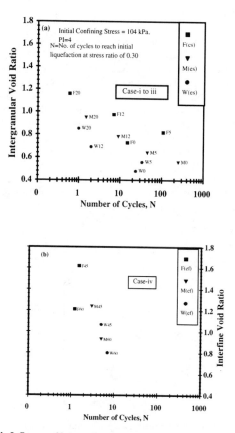

Fig.7a-b Influence of intergrain contacts on cyclic strength: (a) e_c, (b) e_f

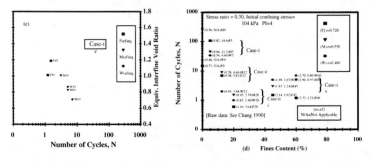

Fig.7c-d Influence of intergrain contacts on cyclic strength: (c) $(e_f)_{eq}$, and (d) FC

Ottawa Sand – Silt Mix

Fig.8 shows the locations of the specimens in the intergranular matrix diagram. Figs.9 and 10a-b show the number of cycles to liquefaction (±5% double amplitudes of axial strain) versus e, e_c and $(e_c)_{eq}$, respectively. In the e versus N plane a distinct trend is observed as FC increases (Fig. 9). At low FC (<FC_{th}) and at the same e, N required to cause liquefaction (at $[\pm\Delta\sigma_1/(2\sigma'_c)]=0.2$) becomes smaller and smaller as the fines content increases. As FC increases beyond FC_{th}, the trend reverses. This is because at low FC, the undrained cyclic strength of the soil is characterized largely by intergranular contact indices (e_c or $(e_c)_{eq}$). Beyond FC_{th}, the interfine contact indices (e_f or $(e_f)_{eq}$) control the undrained cyclic strength. This can be further illustrated when the data are plotted on a e_c versus N or $(e_c)_{eq}$ (assuming b=0.35) versus N plane (Figs. 10a-b). The data points for all soils at FC < FC_{th} collapse into a narrow band in Fig. 10b. This indicates that the undrained cyclic strength of these soils can be characterized by the equivalent intergranular void ratio $(e_c)_{eq}$. Further data are required to evaluate whether the soils at FC > FC_{th} could be characterized by the equivalent interfine void ratio $(e_f)_{eq}$.

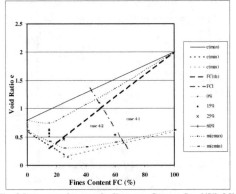

Fig. 8 Intergranular matrix diagram: Ottawa Sand/Silt Mix

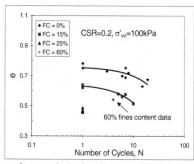

Fig. 9 Effects of *e* on undrained cyclic strength: Ottawa Sand/Silt Mix

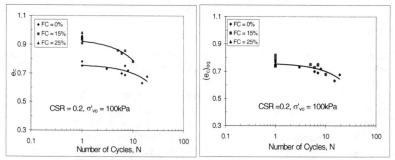

Fig. 10a-b Effects of e_c and $(e_c)_{eq}$ on undrained cyclic strength

Concluding Remarks

A simple framework for analysis of the relative influences of inter-coarser granular and interfine contacts on mechanical response of granular mixes is presented. Equivalent intergranular and interfine contact indices ($(e_c)_{eq}$, and $(e_f)_{eq}$) are introduced as parameters that control undrained cyclic strength of such mixes. With this framework, one may deduce the behavior of granular mixes at various finer grains contents and void ratios relative to the behavior of the host coarser grain soil or finer grain soil. Reinterpretation of some of the existing data from the literature, and interpretation of a limited amount of additional experimental data in light of this framework leads to the following conclusions.

(a) When compared at the same global void ratio e, cyclic strength decreases with an increase in finer grains content up to a certain threshold value FC_{th}. Beyond that the strength increases. FC_{th} depends on the host fines, size disparity ratio, and the global void ratio,

(b) At low FC ($<FC_{th}$), when compared at the same e_c, an increase in finer grains content increases the strength,

(c) At high FC ($>FC_{th}$), when compared at the same e_f, an increase in finer grains content decreases the cyclic strength,

(d) When FC<FC_{th}, equivalent intergranular void ratio $(e_c)_{eq}$ correlates well with cyclic strength of silty sands, and

(e) At high FC (>FC_{th}), $(e_f)_{eq}$ correlates well with cyclic strength of sandy silts.

The above observations do not include effects of possible differences on the nature of field deposits relevant for built environment compared to what is studied in the laboratory.

Acknowledgements

Financial support by the NSF and USGS for the ongoing research is gratefully appreciated.

References

Andrus, R.D. and Stokoe, K.H. (1997). "Liquefaction resistance based on shear wave velocity", Technical Rep. NCEER-97-0022, Highway Project Task No. 112-D-4.2, FHWA Contact No. DTFH61-92-C-00112, 89-128

Baziar, M.H. and Dobry, R. (1995). "Residual strength and large-deformation potential of loose silty sands", J. Geotech. Eng. Div., ASCE, 121(12), 896-906

Chameau, J.L. and Sutterer, K. (1994). "Influence of fines in liquefaction potential and steady state considerations", 13th Intl. Conf., New Delhi, India, 183-84

Chang, N. Y. (1990). "Influence of fines content and plasticity on earthquake-induced soil liquefaction," contract No. DACW3988-C-0078, US Army WES, MS.

Evans, M. D. and Zhou, S. (1995). "Liquefaction behavior of sand-gravel composites", ASCE, J. Geoech. Eng., 121(3), 287-298

Georgiannou, V.N., Burland, J.B. and Hight, D.W. (1990). "The undrained behaviour of clayey sands in triaxial compression and extension", Geotechnique 40(3), 431-449

Georgiannou, V.N., Hight, D.W. and Burland, J.B. (1991). "Undrained behaviour of natural and model clayey sands", Soils and Foundations 31(3), 17-29

Ishihara, K. (1993). "Liquefaction and flow failure during earthquakes", Geotechnique, 43(3), 351-41

JGS (1996). "Soils and Foundations - Special issue on geotechnical aspects of the January 17, 1995 Hyogoken-Nambu Earthquake", Japanese Soc. Geotech. Eng.

Koester, J.P. (1994). "The influence of fines type and conent on cyclic strength", Proc. ASCE Conv., Atlanta, Geotech. Spec. Pub. 44, 17-32

Kuerbis, R., Nagussey, D. and Vaid, Y.P. (1988). "Effect of gradation and fines content on the undrained response of sand", Proc. Conf. Hyd. Fill Struc., ASCE Geot. Spec. Publ. 21, 330-45

Mitchell, J.K. (1993). *Fundamentals of soil behavior,* 2nd Ed., John Wiley & Sons, Inc., New York, N.Y.

Pitman, T.D., Robertson, P.K. and Sego, D.C. (1994). "Influence of fines on the collapse of loose sands", Can. Geotech. J., 31, 728-39

Robertson, P.K and Wride, C.E. (1997). "Cyclic Liquefaction and its Evaluation based on the SPT and CPT", Technical Rep. NCEER-97-0022, Highway Proj. Task No. 112-D-4.2, FHWA Contact No. DTFH61-92-C-00112, 41-87

Roscoe, K. H. (1970). "The influence of strains in soil mechanics." Geotechnique, 20(2), 129-170

Seed, H.B., Idriss, I.M. & Arango, I. (1983). "Evaluation of liquefaction potential using field performance data", J. Geot.. Eng. Div., ASCE, 109, GT3, 458-482

Seed, H.B. (1987). "Design problems in soil liquefaction", J. Geot.. Eng. Div., ASCE, 113(8), 827-45

Seed, R.B., and Harder, L. F. Jr. (1990). "SPT-based analysis of cyclic pore pressure generation and undrained residual strength." Proc., Seed Mem. Symp., 2, 351-376.

Singh, S. (1994). "Liquefaction characteristics of silts." Ground failures under seismic conditions. Proc., ASCE Nat. Convention, Geotech Spec. Publ. 44, S. Prakash and P. Dakoulas, eds., ASCE, Reston, Va., 105-116

Stark, T. D. and Mesri, G. (1992). "Undrained shear strength of liquefied sands for stability analysis", J. of Geotech. Eng., ASCE, 118(11), 1727-47

Thevanayagam, S., Ravishankar, K., and Mohan, S. (1996). "Steady state strength, relative density and fines content relationship for sands," TRB 1547, 61-67

Thevanayagam, S. (1998a) "Effect of fines and confining stress on steady state strength of silty sands," J. Geotech. & Geoenv. Eng., ASCE, 124(6), 479-491

Thevanayagam, S. (1998b). "Relative role of coarser and finer grains on undrained behavior of granular mixes" in review, J. Geotech. & Geoenv. Eng., ASCE.

Thevanayagam, S. (1999). "Role of intergrain contacts, friction, and interactions on undrained response of granular mixes" Proc. workshop Physics and mechanics of soil liquefaction, A.A. Balkema press.

Thevanayagam, S. (1999). "Liquefaction and shear wave velocity characteristics of silty/gravely soils", Proc. 15th Us-Japan Bridge Eng. Workshop, PWRI, Tsukuba City, Japan.

Thevanayagam, S. (2000). "Liquefaction potential and undrained fragility of silty soils", Proc. 12 WCEE 2000 Conf., Auckland, NZ.

Thevanayagam, S. and Mohan, S. (2000). "Intergranular state variables and stress-strain behaviour of silty sands", Geotechnique, 50(1), 1-23.

Vaid, Y.P. (1994). "Liquefaction of silty soils", Proc. ASCE Conv., Geotech. Spec. Publ. 44, 1-16

Yamamuro, J.A. and Lade, P.V. (1998). "Steady-state concepts and static liquefaction of silty sands", J. Geotech. and Geoenv. Engrg. Div., ASCE, 124(9), 868-877.

Zlatovic, S. and Ishihara, K. (1997). "Normalized behavior of very loose non-plastic soils: effects of fabric" Soils and Foundations, 37(4), 47-56

ESTIMATION OF LIQUEFACTION-INDUCED VERTICAL DISPLACEMENTS USING MULTI-LINEAR REGRESSION ANALYSIS

By Mihail Chiru-Danzer[1], C. Hsein. Juang [2] M. ASCE
and R. A. Christopher[3]

ABSTRACT

A new empirical method for evaluating liquefaction-induced vertical displacements is presented, which is based on results obtained from a Multi-Linear Regression (MLR) analysis of field performance records. Previously published techniques for predicting earthquake-induced settlements are based primarily on results derived through laboratory cyclic triaxial testing. Consequently, settlements obtained by such methods account only for the effect of the volumetric strains. In the field, observed vertical displacements are a function not only of the volumetric strains but also the vertical component induced by lateral spreading and the effects of sand boils.

In the present study, a MLR model based on field data is developed for predicting liquefaction-induced vertical displacements. A database consisting of 50 measurements of vertical displacement forms the basis for the MLR model and analysis.

INTRODUCTION

Due to its ease and cost effectiveness, Standard Penetration Testing (SPT) is commonly used to determine the liquefaction potential and the liquefaction-induced displacements in sandy deposits. A wide array of data and publications are available on vibratory compaction of sand and horizontal deformations, but very little work has been published in which liquefaction-induced vertical displacements are evaluated. This paper details a new method of predicting liquefaction-induced vertical displacement using Multi-Linear Regression analysis and is based on a database consisting of 50

---

[1] Project Engineer, LAW-GIBB Group, Inc., 396 Plasters Ave., Atlanta, GA. 30324.
[2] Professor, Clemson University, Dept. of Civil Engineering, Clemson, S.C. 29634.
[3] Associate Professor, Clemson University, Dept. of Geology & Earth Sciences, Clemson, S.C. 29634.

cases. The sources of these cases were the 1964 Niigata, Japan, and the 1971 San Fernando, California, earthquakes. The term, vertical displacement, is used herein to define the total measured displacements that are induced not only by volumetric strains but also result from the vertical component of lateral spreading (Pyke et al. 1975, Darragh et al. 1992) and the effects of sand boils (Darragh et al. 1992).

PAST STUDIES ON SETTLEMENTS OF SAND DUE TO DYANAMIC LOADS

Studies on Vibratory Compaction of Sand

D'Appolonia and D'Appolonia (1967) and D'Applonia, et al. (1969) demonstrated the effects of vibration on sand samples, which were placed in containers at 0% relative density. It was shown that no appreciable settlements occurred within the sample until the loading acceleration approached 100 gals. Furthermore, it was determined that an increase in surcharge stresses yielded a corresponding, proportional increase in the amount of acceleration required to cause particle densification. A similar performance evaluation, Ortigosa and Whitman (1969) concluded that minimal sand densification occurs at accelerations less than 100 gals. It was further shown that for noticeable compression to occur, the dynamic loading must be greater than the static loading generated through overburden pressures. It was concluded that the vertical acceleration alone, induced by earthquakes, has a minimal effect on particle densification, whereas the shear stresses occurring as a result of the horizontal component of acceleration will induce greater subsidence.

Silver and Seed (1971) performed a series of cyclic shear tests, which showed that at a shear strain exceeding 0.05%, the values of vertical stress do not greatly influence the magnitude of settlement as a result of vertical strain. It was also shown that the magnitude of the shear strain directly affects the magnitude of the settlement. Finally, it was concluded that the cyclic shear stresses required for promoting settlement increase with an increase in overburden.

Youd (1971, 1972) performed tests to determine the relative density and compaction of sands through cyclic shear straining. These results confirmed the results discussed previously and demonstrated that settlements generated during simple shear tests are not dependent on the loading frequency or sample saturation.

Seed and Silver (1972) proposed a procedure for estimating settlement in unsaturated sands based on a response analysis of the soil stratum to base excitation in order to determine the induced shear strains generated at various depths within the sand stratum. This procedure was based on the assumption that the movement within the sand stratum may be approximated through vertically propagating shear waves. Pyke et al. (1975) modified the Seed and Silver (1972) procedure to allow for multidirectional shaking effects, which were shown to have a major impact on the settlement magnitude. It was shown that the effect generated by combined horizontal motion is approximately equal to the summation of the effects of the individual components. It was concluded that multidirectional shaking produced settlements three-fold larger than those generated during one-directional shaking.

Drabkin et al. (1997) used regression analysis to demonstrate that maximum densification of sand occurs when high deviator stresses are combined with low confining pressures. Other related studies were performed by Ishihara and Yasuda (1972); Peacock and Seed (1968); Seed and Idriss (1971); Seed and Lee (1966); Yoshimi (1967); Chen (1988); Darragh et al. (1992), and Baziar et al. (1992). More recently, Kim et al. (1994) and Kim and Drabkin (1995) performed laboratory analyses of low-level excitation on sands. Their results show that settlement depends on conditions of stress, intrinsic sand properties in drained conditions, and vibration parameters.

Studies on Earthquake-Induced Compaction

Lee and Albaisa (1974) investigated earthquake-induced settlements in saturated sands with no capacity for drainage, and the subsequent settlement resulting from the dissipation of the pore water pressures. Tatsuoka et al. (1984) studied volumetric strain after initial liquefaction and concluded that settlement is sensitive to the amount of maximum shear strain developed and to the soil density, but not to the effective overburden pressure.

Tokimatsu and Seed (1987) proposed a simplified method for evaluating settlements in sands due to earthquake shaking. The simplified method proposed therein estimates probable settlements in both saturated and unsaturated sands, and considers the effects of the cyclic stress ratio in saturated sands and the cyclic shear strain in partially saturated or dry sands. Other parameters considered in their evaluation are the SPT N-values together with the magnitude and maximum ground acceleration of the earthquake. It should be noted that the SPT N-values are standardized for Japanese standards, effective overburden, and a drill rig operating at 60% efficiency.

Ishihara and Yoshimine (1991) proposed another method for determining earthquake induced settlements that includes both SPT and Cone Penetration Test (CPT) data. A graphic representation was subsequently derived relating the SPT N-values and cone tip resistance to shear and volumetric strains.

DEVELOPMENT OF NEW EMPIRICAL MODELS

Introduction

Multiple linear regression (MLR) is often used in geotechnical engineering to establish empirical equations for predicting the behavior of complex phenomena. MLR operates on the premise that the changes within the independent variable X are accompanied by changes in the dependent variable Y. A multiple linear regression model takes the form:

$$Y^* = \beta_0 + \beta_1 X_1 + \beta_2 X_2 + \beta_3 X_3 + \ldots + \beta_n X_n + \varepsilon \tag{1}$$

where

Y^* = Predicted dependent variable,

$\beta_0 \ldots \beta_n$ = Partial regression coefficients,

$X_0 \ldots X_n$ = Independent variables, and

ε = $Y - Y^*$ = the difference between the observed value of the dependent variable and the predicted value of the dependent variable.

The degree to which the MLR model accounts for the total variability in the dependent variable is gauged by the coefficient of determination, R^2, which is defined as:

$$R^2 = \frac{S_t - S_r}{S_t} \tag{2}$$

where

R^2 = Coefficient of determination, which is a measure of the variability of the dependent variable Y with respect to changes in the independent variable X,

S_t = $\sum (Y_i - \bar{Y})^2$

S_r = $\sum (Y_i - \hat{Y}_i)^2$

Y_i = Measured dependent variable,

Y^* = Predicted dependent variable,

$\bar{Y}$ = Mean of the measured dependent variable.

In the present study, an empirical model is developed using a stepwise multiple linear regression procedure. With this procedure, independent variables are entered into the model one at a time; the variable selected for inclusion in the model is the one that accounts for the greatest amount of variation in 'Y' not accounted for by the variables included in previous steps. In each iteration, all partial regression coefficients are tested against the user-selected significance level, with the result that variables included in earlier steps may be eliminated from the model. Bartlett and Youd (1995) published a well-documented manuscript presenting their MLR approach to liquefaction-induced lateral spreads.

Source Data

The data used in the MLR analysis was obtained from two primary sources. Locations 1 through 40 were obtained from the 1964 Niigata, Japan, Earthquake. Table 1 presents the sources and references for these locations. All Niigata data was extrapolated from working maps developed by Hamada et al. (1992). Furthermore, locations 41 through 50 were obtained from the vicinity of the Joseph Jensen Filtration Plant, a site that had liquefied during the 1971 San Fernando, California, Earthquake. The San Fernando data was obtained from working maps developed by O'Rourke et al. (1992).

Table 1. Performance of Regression Models And Overall Effect of Input Variables In Predicting Vertical Displacement 'h'

Data Number	Thickness of Liquefied Layer (ft)	Slope of liquefied layer (%)	N	Amax (g)	Mw	Seismic load	Distance from nearest water source (ft)	Vertical Displ. (ft)	Earthquake	Year	Site	References
1	16.4	4.0	9.2	.25	7.5	.25	49	4.30	Niigata	1964	Bandai to Yachiyo Br.	Hamada et al. (1992)
2	13.1	4.0	9.5	.25	7.5	.25	82	2.76	Niigata	1964	Bandai to Yachiyo Br.	Hamada et al. (1992)
3	42.7	2.0	9.8	.25	7.5	.25	82	7.35	Niigata	1964	Bandai to Yachiyo Br.	Hamada et al. (1992)
4	21.3	12.0	17.3	.25	7.5	.25	394	4.43	Niigata	1964	Bandai to Yachiyo Br.	Hamada et al. (1992)
5	42.7	4.0	9.8	.25	7.5	.25	164	6.56	Niigata	1964	Bandai to Yachiyo Br.	Hamada et al. (1992)
6	42.7	4.0	9.8	.25	7.5	.25	164	6.63	Niigata	1964	Bandai to Yachiyo Br.	Hamada et al. (1992)
7	42.7	4.0	9.8	.25	7.5	.25	98	5.32	Niigata	1964	Bandai to Yachiyo Br.	Hamada et al. (1992)
8	42.7	4.0	9.8	.25	7.5	.25	131	6.56	Niigata	1964	Bandai to Yachiyo Br.	Hamada et al. (1992)
9	4.9	6.0	14.9	.25	7.5	.25	82	.43	Niigata	1964	Yachiyo to Showa Br.	Hamada et al. (1992)
10	36.1	5.0	5.9	.25	7.5	.25	164	8.89	Niigata	1964	Yachiyo to Showa Br.	Hamada et al. (1992)
11	32.8	5.0	6.1	.25	7.5	.25	131	7.12	Niigata	1964	Yachiyo to Showa Br.	Hamada et al. (1992)
12	26.2	5.0	9.9	.25	7.5	.25	230	6.07	Niigata	1964	Yachiyo to Showa Br.	Hamada et al. (1992)
13	13.1	3.0	19.1	.25	7.5	.25	361	2.82	Niigata	1964	Yachiyo to Showa Br.	Hamada et al. (1992)
14	23.0	2.5	15.3	.25	7.5	.25	1476	3.94	Niigata	1964	Kawagishi-cho	Hamada et al. (1992)
15	23.0	2.5	15.3	.25	7.5	.25	1476	4.07	Niigata	1964	Kawagishi-cho	Hamada et al. (1992)
16	16.4	6.0	18.4	.25	7.5	.25	1312	2.79	Niigata	1964	Kawagishi-cho	Hamada et al. (1992)
17	16.4	6.0	18.4	.25	7.5	.25	1312	2.72	Niigata	1964	Kawagishi-cho	Hamada et al. (1992)
18	16.4	6.0	18.4	.25	7.5	.25	1312	3.05	Niigata	1964	Kawagishi-cho	Hamada et al. (1992)
19	36.1	3.0	14.9	.25	7.5	.25	82	4.63	Niigata	1964	Kawagishi-cho	Hamada et al. (1992)
20	39.4	1.0	7.2	.25	7.5	.25	246	4.56	Niigata	1964	Bandai to Yanagishima	Hamada et al. (1992)
21	23.0	2.0	11.9	.25	7.5	.25	574	1.64	Niigata	1964	Estuary of Shinano Rvr.	Hamada et al. (1992)
22	42.7	9.0	12.6	.25	7.5	.25	164	9.71	Niigata	1964	Bandai Island	Hamada et al. (1992)
23	42.7	4.0	12.6	.25	7.5	.25	82	8.10	Niigata	1964	Bandai Island	Hamada et al. (1992)
24	29.5	9.0	14.3	.25	7.5	.25	394	6.59	Niigata	1964	Bandai Island	Hamada et al. (1992)

Table 1 (Cont'd). Performance of Regression Models And Overall Effect of Input Variables In Predicting Vertical Displacement 'h'

25	27.9	3.5	19.4	.25	7.5	.25	82	3.77	Niigata	1964	Fishery Pier	Hamada et al. (1992)
26	9.8	2.0	7.7	.25	7.5	.25	984	.20	Niigata	1964	Hotel Niigata	Hamada et al. (1992)
27	13.1	2.0	7.5	.25	7.5	.25	1181	.75	Niigata	1964	Hotel Niigata	Hamada et al. (1992)
28	26.2	3.0	9.3	.25	7.5	.25	738	4.66	Niigata	1964	Hotel Niigata	Hamada et al. (1992)
29	29.5	1.0	14.2	0.25	7.5	0.253024	1476	1.9	Niigata	1964	Hotel Niigata	Hamada et al. (1992)
30	29.5	1.0	14.2	.25	7.5	.25	1394	2.62	Niigata	1964	Hotel Niigata	Hamada et al. (1992)
31	29.5	1.0	14.2	.25	7.5	.25	1312	2.95	Niigata	1964	Hotel Niigata	Hamada et al. (1992)
32	13.1	4.0	14.9	.25	7.5	.25	6562	3.97	Niigata	1964	Ebigase	Hamada et al. (1992)
33	13.1	4.0	14.9	.25	7.5	.25	6562	3.35	Niigata	1964	Ebigase	Hamada et al. (1992)
34	16.4	4.0	11.6	.25	7.5	.25	6562	4.76	Niigata	1964	Ebigase	Hamada et al. (1992)
35	19.7	4.0	11.2	.25	7.5	.25	6562	5.12	Niigata	1964	Ebigase	Hamada et al. (1992)
36	21.3	0.5	9.7	.25	7.5	.25	1641	1.48	Niigata	1964	Shitayama	Hamada et al. (1992)
37	9.8	1.0	13.9	.25	7.5	.25	1641	1.87	Niigata	1964	Shitayama	Hamada et al. (1992)
38	11.5	1.0	13.7	.25	7.5	.25	1641	1.90	Niigata	1964	Shitayama	Hamada et al. (1992)
39	16.4	1.0	11.6	.25	7.5	.25	1641	2.99	Niigata	1964	Shitayama	Hamada et al. (1992)
40	13.1	1.0	11.9	.25	7.5	.25	1805	1.48	Niigata	1964	Shitayama	Hamada et al. (1992)
42	23.0	1.5	21.6	.75	6.4	.57	1312	.43	San Fernando	1971	Joseph Jensen Filtration Plant	O'Rourke et al. (1992)
43	8.2	1.0	9.8	.75	6.4	.57	1312	.43	San Fernando	1971	Joseph Jensen Filtration Plant	O'Rourke et al. (1992)
44	6.6	1.0	9.8	.75	6.4	.57	1312	.30	San Fernando	1971	Joseph Jensen Filtration Plant	O'Rourke et al. (1992)
45	6.6	1.0	9.8	.75	6.4	.57	1312	.26	San Fernando	1971	Joseph Jensen Filtration Plant	O'Rourke et al. (1992)
46	6.6	1.0	9.8	.75	6.4	.57	1312	.26	San Fernando	1971	Joseph Jensen Filtration Plant	O'Rourke et al. (1992)
47	26.2	0.5	17.6	.75	6.4	.57	1312	.79	San Fernando	1971	Joseph Jensen Filtration Plant	O'Rourke et al. (1992)
48	9.8	6.0	11.8	.75	6.4	.57	1312	2.99	San Fernando	1971	Joseph Jensen Filtration Plant	O'Rourke et al. (1992)
49	6.6	1.5	8.8	.75	6.4	.57	1312	.66	San Fernando	1971	Joseph Jensen Filtration Plant	O'Rourke et al. (1992)
50	6.6	1.5	8.8	.75	6.4	.57	1312	.69	San Fernando	1971	Joseph Jensen Filtration Plant	O'Rourke et al. (1992)

The accuracy of the MLR model depends on the quality of the input. It should be noted that all deformations generated by the 1964 Niigata and 1971 San Fernando earthquakes were measured via aerial photographs that have an inherently large margin of error. Hamada et al. (1992) reported the accuracy of the measurement in the vertical direction on the order of 0.4 to 0.5 meters. In an effort to reduce the amount of error, the writers used only data available within 10 meters of the nearest shown profile. If great spatial variability was evident, the data used was restricted to 5 meters of the nearest cross-sectional profile.

Multi-Linear Regression (MLR) Model

The sensitivity analysis (Chiru-Danzer, 1999) performed on the artificial neural network (ANN) vertical displacement models has shown that some variables affect liquefaction-induced vertical displacements to a greater degree than do others. The thickness of the liquefied layer, the SPT, $(N_1)_{60}$-values, and the seismic load parameter are critical factors that must be included into any model that predicts vertical displacement.

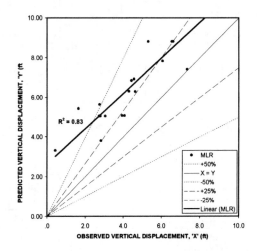

Figure 1. Comparison of Actual vs. Predicted Values for Model MLR

The first regression model is based on 21 data points (only cases that were located on the cross-sectional profiles shown in Hamada et al., 1992) collected solely from the 1964 Niigata Earthquake. Many observed settlements were reported in Hamada (1992); however, very few of the variables required to model the magnitude of the vertical displacement were available, based on the cross-sectional profiles. The greater gradient of the ground surface or the lower boundary of the liquefied layer, θ (after Hamada, 1992), was obtained based on the cross-sectional profiles and is consequently as accurate as the information was presented. To minimize error associated with collecting data far from the profiles, only data within five to ten meters from the cross section was used to develop the subsequent models. The following model was derived using multi-linear regression:

$$\Delta h = [1-\log_{10}(N^{0.56})] \bullet (\theta^{0.25}) \bullet (T^{0.5})$$ (3)

where

Δh	=	Vertical displacement (ft),
N	=	Corrected SPT blow count (bpf)
θ	=	The greater gradient of ground surface or lower boundary of the liquefied layer (%) (after Hamada, 1992),
T	=	Thickness of the liquefied layer (ft).

The comparison of the vertical displacements of the 21 cases predicted by Equation (3) with the observed displacements are shown in Figure 1. Equation (3) generally over-estimates the vertical displacement, as is evident from Figure 1. However, a strong correlation exists between the predicted and observed displacements, as evident by a R^2 of 0.83. About 86% of the predictions fall within ±50% of the measured vertical displacements.

<u>Stepwise Multi-Linear Regression (SMLR) Models</u>

Influence of Thickness of Liquefied Layer

To better understand the behavior of the individual input variables upon the overall model, stepwise regression was chosen to determine the best model. All data used in Table 1 were used in the analysis.

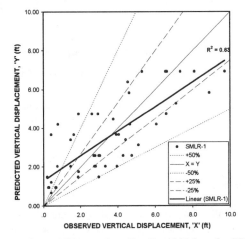

Figure 2. Comparison of Observed vs. Predicted Values for Model SMLR-1

Using the software package SPSS (1999), the first variable added was the thickness of the liquefied layer, yielding the following equation:

$$\Delta h = -0.13 + 0.164 \ (T) \tag{4}$$

The thickness was entered as the first variable due to the highest level of correlation between it and the dependent variable Δh. The R^2 for this model (SMLR-1) is 0.634, as shown in Figure 2.

Influence of Ground Slope

Due to the importance of the greater slope between the ground surface and the lower boundary of the liquefied layer θ, the ground slope was selected by the stepwise procedure as the second variable to be entered into the equation. Its introduction to the model considerably improved the SMLR model.

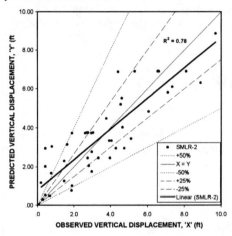

Figure 3. Comparison of Observed vs. Predicted Values for Model SMLR-2

The coefficient of determination for the model at this step of the regression, referred to herein as Model SMLR-2, was 0.78, as shown in Figure 3. The new equation obtained was:

$$\Delta h = -1.082 + 0.147 \ (T) + 0.401 \ (\theta) \tag{5}$$

Influence of SPT $(N_1)_{60}$-Value

The third variable added to the stepwise model is the Standard Penetration Resistance $(N_1)_{60}$-value, which again improved the R^2 value of the final model to 0.82. With the inclusion of the SPT $(N_1)_{60}$-value, the SMLR model becomes:

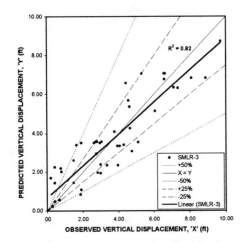

Figure 4. Comparison of Observed vs. Predicted Values for Model SMLR-3

$$\Delta h = 0.57 + 0.141\ (T) + 0.452\ (\theta) - 0.136\ (N) \tag{6}$$

The comparison between the measured and predicted values for this model, referred to herein as Model SMLR-3, is illustrated in Figure 4.

Influence of Distance from Nearest Water Source

A decrease in the distance to the nearest water source will yield, in most cases, a potential for higher horizontal displacements (Hamada, 1992). Because the vertical displacements predicted by these models are predominantly a function of the vertical component of lateral spreading, the distance to the nearest water source D is associated with the dependent variable Δh. The model obtained after the introduction of the variable D is:

$$\Delta h = 0.0439 + 0.154\ (T) + 0.461\ (\theta) - 0.151\ (N) + 0.000257\ (D) \tag{7}$$

The R^2 for this model, referred to herein as Model SMLR-4, is 0.85. The comparison between the measured and predicted values is illustrated in Figure 5.

Influence of Seismic Load

The other variables considered were the earthquake magnitude, M_w and the maximum ground acceleration, a_{max}. In the present study, a seismic load (SL) parameter, defined by (Chen, 1999; Juang et al., 2000), is used:

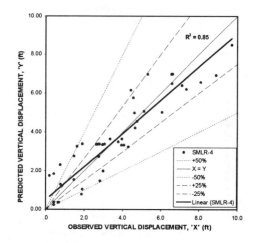

Figure 5. Comparison of Observed vs. Predicted Values for Model SMLR-4

$$SL = a_{max}/MSF \tag{8}$$

where MSF is the magnitude scaling factor defined below (Idriss, 1999):

$$MSF = 37.9 \bullet M_W^{-1.81} \quad \text{for } M_W \geq 5.75 \text{ and} \tag{9}$$

$$MSF = 1.625 \qquad \text{for } M_W < 5.75 \tag{10}$$

With the addition of the variable SL, the following equation is obtained:

$$\Delta h = 0.943 + 0.145(T) + 0.439\,(\theta) - 0.151\,(N) + 0.00026\,(D) - 1.871\,(SL) \tag{11}$$

The final and highest R^2 obtained for this model, referred to herein as Model SMLR-5, is 0.854. The plot of the measured versus predicted values of vertical displacement is presented in Figure 6. Addition of the variable SL does not improve significantly the accuracy of the resulting model. This might be expected, as all data used in the analysis came from only two earthquakes.

A summary of the performance of the above regression models is presented in Table 2. In addition to R^2 values, the success rates, defined at two levels of accuracy, are shown. Model SMLR-5 is shown to be the best among all six models examined.

Plots of the residuals are presented in Figure 7 ('a' through 'e') for each of the input variables, respectively. The residuals plotted on the 'Y' axes are the difference between

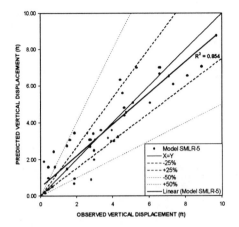

Figure 6. Comparison of Observed vs. Predicted Values for Model SMLR-5

the observed and the predicted values. The independent variables, thickness of the liquefied layer, ground slope, N-value, distance to the nearest water source, and seismic load, are plotted on the 'X' axes. A homoscedastic association is apparent in all of the scatter plots, since there is a random distribution of the residuals about zero.

Table 2. Performance of Regression Models for Predicting Vertical Displacement 'Δh'

Model	Inputs	R^2 for measured vs. predicted displacement	Success Rate (within ± 25% of measured value)	Success Rate (within ± 50% of measured value).
MLR	$S = f(\log(N), \theta, T)$	0.83	30%	60%
SMLR-1	$S = f(T)$	0.63	38%	78%
SMLR-2	$S = f(T, \theta)$	0.78	48%	86%
SMLR-3	$S = f(T, \theta, N)$	0.82	66%	86%
SMLR-4	$S = f(T, \theta, N, D)$	0.85	70%	90%
SMLR-5	$S = f(T, \theta, N, D, SL)$	0.854	76%	90%

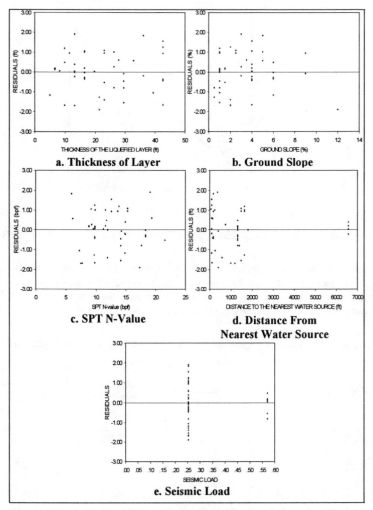

Figure 7 (a-e). Residual Plots of Input Variables For Regression Model 2

The residual plot of the independent variable, seismic load is presented in Figure 7e. Because the data were obtained from two earthquake sites (Niigata, 1964 and San Fernando, 1971), only one value of seismic load was obtained at each of the sites. Consequently, there are only two values along the 'X' axis where the distribution of residuals about zero is significant. Nonetheless, there is an approximately even distribution of residuals about the 'zero' intercept, which indicates that this relationship also is homoscedastic, and should therefore provide adequate results in future predictions.

It is recommended that Equation (11) (Model SMLR-5) be used with the following ranges of input variables:

T	= 0 to 50 (ft),
θ	= 0 to 12 (%),
N	= 0 to 30 (bpf),
D	= 0 to 1500 (ft), and
SL	= 0.20 to 0.6.

Equation (11) may be used to predict the total amount of vertical displacement at sites where the expected lateral spreads are potentially large (i.e., sites with a large free face ratio, ground slope, and/or near a water source).

CONCLUSION

The empirical MLR model presented in Equation (11) provides a good predictive capability ($R^2 = 0.854$) and success rate (90% of the cases fall within $\pm$ 50% of the observed values) based on the limited data examined. Further validation of the developed MLR model using additional database, particularly using data from earthquakes with various SL values, is warranted.

REFERENCES:

Bartlett, S.F., and Youd, T.L., (1995), "Empirical Prediction of Liquefaction-Induced Lateral Spread," ASCE Journal of Geotechnical Engineering, Vol. 121, No. 4., April, pp. 316-328.

Baziar, M.H., Dobry, R., and Elgamel, A.W.M. (1992), "Engineering evaluation of permanent ground deformations due to seismically-induced liquefaction," *Technical Report NCEER-92-0007*, National Center for Earthquake Engineering Research, State University of New York, Buffalo.

Chen, A.T.F., (1988), "Analysis of Strong Ground Motions and Associated large Deformations on Seismically Induced Pore Pressure and Settlement," *Earthquake Engineering and Soil Dynamics II – Recent Advances in Ground-Motion Evaluation*, Proceedings of the specialty conference sponsored by the Geotechnical Engineering Division of the ASCE, Special Publication No. 20, pp. 482-492.

Chen, J.C. (1999), "Risk-Based Liquefaction Potential Evaluation Using Cone Penetration Tests and Shear Wave Velocity Measurements," *Ph.D. Dissertation*, Clemson University, Clemson, SC 29630.

Chiru-Danzer (1999), "Estimation of Liquefaction-Induced Vertical and Horizontal Displacements Using Artificial Neural Networks and Regression Analysis" *Ph.D. Dissertation*, Clemson University, Clemson, SC 29630.

D'Appolonia, D.J., and D'Appolonia, E. (1967), "Determination of the Maximum Density of Cohesionless Soils," *Proceedings of the Third Asian Conference on Soil Mechanics and Foundation Engineering*, Vol. 1, Haifa, Israel, , pp. 266-268.

D'Appolonia, D.J., Whitman, R.V., and D'Appolonia, E. (1969), "Sand Compaction with Vibration Rollers," *Journal of Soil Mechanics and Foundation Division*, ASCE, Vol. 95, No. SM1, Proc. Paper 6366, pp. 263-284.

Darragh, R., Taylor, H.T., Scawthorn, C., Seidel, D., and Ng, C. (1992), "Liquefaction Study , Sullivan Marsh and Mission Creek, San Francisco, California," *Proceedings from the Fourth Japan-U.S. Workshop on Earthquake Resistant Design of Lifeline Facilities and Countermearsures for Soil Liquefaction, Technical Report NCEER-92-0019*, National Center for Earthquake Engineering Research, Buffalo, NY, Vol.1, pp. 205-222.

Drabkin, S., Lacy, H., and Kim, D.S. (1997), "Estimating Settlement of Sand Caused by Construction Vibration," *Journal of Geotechnical Engineering*, ASCE, Vol. 122, No. 11, pp. 920-928.

Hamada, M., and O'Rourke, T.D., Eds., (1992), "Large Ground Deformations and Their Effects on Lifelines: 1979 San Fernando Earthquake," *Case Studies of Liquefaction and Lifeline Performance During Past Earthquakes, Technical Report NCEER-92-0001*, National Center for Earthquake Engineering Research, Buffalo, NY, Vol.1.

Idriss, I.M. (1999), An update of the Seed-Idriss simplified procedure for evaluating liquefaction potential, Proceedings, TRB Workshop on New Approaches to Liquefaction Analysis, FHWA-RD-99-165, Federal highway Administration, Washington, D.C.

Ishihara, K., and Yasuda, S. (1972), "Sand Liquefaction Due to Irregular Excitation," *Soils and Foundations*, Vol. 12, No. 4, pp. 65-77.

Ishihara, K., and Yoshimine, M. (1992), "Evaluation of Settlements in Sand Deposits Following Liquefaction During Earthquakes," *Soils and Foundations*, Vol. 32, No.1, pp. 173-188.

Juang, C.H., Chen, C.J., Rosowsky, D.V., and Tang, W.H. (2000), "A Risk-based Method for Assessing Liquefaction Potential Using CPT," Geotechnique, The Institution of Civil Engineers, U.K., (accepted for publication).

Kim, D.S., Drabkin, S., Laefer, D., Rokhvarger, A., (1994), "Prediction of low level vibration induced settlement," Vertical and Horizontal Deformation of Foundation and Embankments, *Proc. of Settlement '94*, A.T. Yeung and G.Y. Felio, eds., ASCE, New York, N.Y., pp. 806-817.

Kim, D.S., and Drabkin, S. (1995), "Investigation of vibration induced settlement using multifactorial experimental design," *Geotech. Testing J.*, 18(4), pp. 463-471.

Lee, K.L., and Albaisa, A. (1974), "Earthquakes Induced Settlements in Saturated Sands," *Journal of Geotechnical Engineering*, ASCE, Vol. 100, No. GT7, pp. 397-406.

O'Rourke, T.D., Eds., and Hamada, M., (1992), "Large Ground Deformations and Their Effects on Lifelines: 1964 Niigata Earthquake," *Case Studies of Liquefaction and Lifeline Performance During Past Earthquakes, Technical Report NCEER-92-0002*, National Center for Earthquake Engineering Research, Buffalo, NY, Vol.2.

Ortigosa, P., and Whitman, R.V. (1969), "Densification of Sand by Vertical Vibrations with almost constant stresses," *Soils Publication No. 206*, Massachusetts Institute of Technology, Cambridge, Mass.

Peacock, W.H., and Seed, H.B. (1968), "Sand Liquefaction Under Cyclic Loading Simple Shear Conditions," *Journal of the Soil Mechanics and Foundations Division*, ASCE, Vol. 94, No. SM3, pp. 689-708.

Pyke, R., Seed, H.B., Chan, C.K (1975), "Settlement of Sands Under Multidirectional Shaking," *Journal of the Geotechnical Engineering Division*, ASCE, Vol. 101, No. GT4, pp. 379-398.

Seed, H.B., and Idriss, I.M. (1971), "Simplified Procedure for Evaluating Soil Liquefaction Potential," *Journal of the Soil Mechanics and Foundations Division*, ASCE, Vol. 97, No. 9, pp. 1249-1273.

Seed, H.B. and Lee, K.L. (1966), "Liquefaction of Saturated Sands During Cyclic Loading," *Journal of the Soil Mechanics and Foundations Division*, ASCE, Vol. 92, No. SM6, pp. 105-134.

Seed, H.B., and Silver, M.L (1972), "Settlement of Dry Sands During Earthquakes," *Journal of Soil Mechanics and Foundation Division*, ASCE, Vol. 93, No. SM3, pp. 381-397.

Silver, M.L., and Seed, H.B. (1971), "Volume Change in Sands During Cyclic Loading," *Journal of Soil Mechanics and Foundation Division*, ASCE, Vol. 97, No. SM9, pp. 1081-1098.

SPSS 9.0 (1999), SPSS Base 9.0 Applications Guide and User's Guide, SPSS, Inc.

Tatsuoka, F., Sasaki, T., and Yamada, S. (1984), "Settlement in saturated sand induced by cyclic undrained simple shear," *Eighth World Conference on Earthquake Engineering*, San Francisco, Vol. III, pp. 95-102.

Tokimatsu, K., and Seed, H.B. (1987), "Evaluation of Settlements in Sands due to Earthquake Shaking," *Journal of Geotechnical Engineering*, ASCE, Vol. 113, No. 8, pp. 861-878.

Yoshimi, Y. (1967), "An Experimental Study of Liquefaction of Saturated Sands," *Soils and Foundations*, Vol. 7, No. 2, pp. 20-32.

Youd, T.L. (1971), "Maximum Density of Sand by Repeated Straining in Simple Shear," *Highway Research Record*, No. 374, pp. 1-6.

Youd, T.L. (1972), "Compaction of Sands by Repeated Shear Straining," *Journal of the Soil Mechanics and Foundations Division*, ASCE, Vol. 98, No. SM7, pp. 709-725.

POST-LIQUEFACTION FLOW DEFORMATIONS

W.D. Liam Finn[1]

ABSTRACT

The state of the art for evaluating post liquefaction flow deformations of earth structures is presented and the main parameters that affect the deformations are reviewed. The primary focus is on the estimation of residual strength and large displacement methods for analyzing the consequences of liquefaction.

INTRODUCTION

One of the most challenging problems facing geotechnical engineers is the seismic safety evaluation of soil structures such as embankment dams, which have potentially liquefiable soils in the structure itself or in the foundation. This problem poses three difficult questions;

- Will liquefaction be triggered
- If so what are the consequences
- What remediation measures should be adopted ensure satisfactory behaviour.

The triggering of liquefaction was reviewed in 1996 by a Committee appointed by the National Center for Earthquake Engineering at the University of Buffalo. The committee reviewed the state of the art of practice and research since a similar review was conducted in 1985 (NRC, 1985) and made several recommendations for improving the state of practice. The recommendations of the committee and the supporting documentation are reported in NCEER (1997). In view of this comprehensive study, the triggering of liquefaction will not be reviewed in this paper. Here the focus will be on issues related to evaluating the consequences of liquefaction.

---

[1]Anabuki Chair of Foundation Geodynamics, Kagawa University, Japan, and Professor Emeritus, University of British Columbia, Vancouver, B.C., Canada

In the context of this review, liquefaction is synonymous with strain softening of relatively loose sands in undrained shear as illustrated by curve 1 in Fig. 1. When the sand is strained beyond the point of peak strength, the undrained strength drops to a value that is maintained constant over a large range in strain. This is conventionally called the undrained steady state or residual strength. If the strength increases after passing through a minimum value, the phenomenon is called limited or quasi-liquefaction and is illustrated by curve 2 in Fig. 1. Even limited liquefaction may result in significant deformations because of the strains necessary to develop the strength to restore stability.

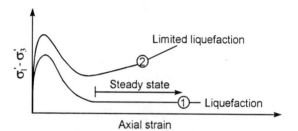

Fig. 1. Types of contractive deformation (Vaid et al., 1989).

The appropriate residual strength for design and analysis is a very controversial matter. This was clearly evident from the proceedings of a major workshop on the shear strength of liquefied soils which was held at the University of Illinois at Urbana in 1997 with broad representation from the research and engineering community (NSF, 1997) and of an international workshop at Johns Hopkins University in Baltimore in 1998 with emphasis on the physics and mechanics of soil liquefaction (Lade and Yamamuro, 1998). The two workshops make very significant contributions to understanding what controls the residual strength of soils. But they also demonstrate the sometimes widely divergent opinions that exist on even the most basic issues.

The focus of this paper is on engineering practice. Many interesting theoretical issues are ignored. They can be found in the proceedings of the two workshops cited above. The primary objective here is to provide a coherent framework of understanding of research findings and methods of analysis to the practicing engineer.

Very few field data have been available to validate our methods for analysing the consequences of liquefaction. The 1993 Kushiro and the 1994 Nansei earthquakes in Hokkaido, Japan caused widespread damage to flood protection dikes by liquefaction and provided a data base for validation of large displacement finite element analysis of post-liquefaction displacements. The results of a validation study for the Hokkaido Development Bureau will be described.

RESIDUAL STRENGTH FROM LABORATORY TESTS

Practice before 1988

Until the late 1980's, residual strength was determined using undrained triaxial compression tests on undisturbed samples from the field or on samples reconstituted to the field void ratio using moist tamping. This approach followed from the pioneering work of Castro (1969). Potentially liquefiable soils are very difficult to sample without disturbance. They are likely to densify during sampling, transportation, and during the process of setting up the samples for testing. Therefore, tests cannot be conducted at the field void ratio. Since the residual strength was considered to be a function of the void ratio only, a logical solution to the disturbance problem was to correct the laboratory residual strength for the effects of changes in void ratio. Poulos et al. (1985) developed such a procedure. However the corrections for disturbance can lead to order of magnitude changes in the measured residual strength. Such large corrections are a matter of concern.

The consequences of liquefaction were assessed primarily by limiting equilibrium analyses of stability. Levels of safety and remediation requirements were defined in terms of acceptable factors of safety. In some instances, displacement criteria were also used in addition to factors of safety. Displacements were estimated using the Newmark sliding block method of analysis. In applying this method, the residual strength was used in determining the yield acceleration. The Newmark method is not an appropriate method for analyzing structures with large volumes of liquefied material undergoing complicated internal distortions. It is best left for those situations envisaged by Newmark in which displacements are constrained to relatively narrow zones of concentrated shear. Large displacement finite element analysis is now being used in practice to determine post-liquefaction deformations in embankment dams. About 15 dams have been analysed in this way since 1989. This type of analysis will be discussed later.

Practice after 1988

In 1987, Harry Seed published the results of a study that changed drastically the state of practice (Seed, 1987). He determined representative values of residual strength by back-analyzing embankments which had undergone significant displacements during earthquakes. The materials yielding these strengths were characterized by corrected, normalized Standard Penetration Resistances, $(N_1)_{60}$. An updated version of his original correlation chart, developed by Seed and Harder (1990), is shown in Fig. 2.

There is no data beyond $(N_1)_{60}$ of 15. However the curve is often extrapolated beyond this range to provide values of residual strength at higher penetrations for safety evaluation and remediation studies. There is considerable scatter in the data and in the region near the lower bound the residual strengths are negligible for $(N_1)_{60}$

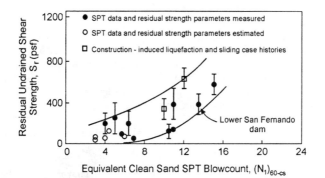

Fig. 2. Relationship between corrected "clean sand" blowcount $(N_1)_{60c-s}$ and undrained residual strength, s_r, from case studies (Seed and Harder, 1990).

less than 12. Most designers opted for strengths between the lower bound and the 33 percentile. Since these values were generally substantially less than would be given by triaxial compression tests, there was a considerable impact on seismic safety assessments and the extent of required remediation.

A re-evaluation of the liquefaction induced failure of the San Fernando dam which occurred during the 1971 San Fernando earthquake was undertaken by both Castro and Seed in 1986-1987 with the objective of resolving the uncertainties surrounding the determination of residual strength. Seed et al. (1989) reported that the average steady state strength of all samples tested in undrained compression was 5250 psf before correction for disturbance, and 800 psf after correction, a correction factor of about 6.5. The corrected average value did not allow the dam to fail by sliding instability in a static equilibrium analysis. The average residual strength obtained from back-analysis of the failed dam in the final configuration was 400±100 psf. The 35 percentile residual strength based on laboratory data would predict failure of the San Fernando dam. This suggests that, on the average, laboratory compression tests overestimate the residual strength. The San Fernando study did not resolve all the difficulties surrounding the determination of residual strength. However, use of the Seed (1987) chart for estimating reduced strength became widespread in engineering practice after this study.

FACTORS CONTROLLING RESIDUAL STRENGTH

Stress Path

Vaid and Chern (1985) showed that the residual strength measured in extension was much smaller than the strength in compression and that sands in a given state were much more contractive in extension than in compression. These differences are often dismissed as being due to non-uniformity of the test specimens and the development of necking at large strains in extension tests. However, in the study by Vaid and Chern (1985), samples were frozen both before and after undrained extension tests by injecting gelatin following the procedures described by Emery et al. (1973). The samples were then sliced to determine the distributions of void ratio before and after testing. In these samples, collapse occurred at 5% strain and loading was continued to 9% strain. The samples showed remarkable uniformity, both before and after testing. Tests by Vaid and Sivathayalan (1996), Reimer et al. (1997) and Yoshimine et al. (1998), have confirmed these results for triaxial tests and have also shown that the strength in simple shear was significantly less than in triaxial compression.

Uthayakumar and Vaid (1998) and Yoshimine et al. (1998) have explored the effects of stress path on residual strength over a wide range of stress paths defined by α, the inclination of the principle stress to the vertical axes of the sample, and the parameter $b = (\sigma_2 - \sigma_3)/(\sigma_1 - \sigma_3)$ which is a measure of the intermediate principle stress.

The samples were tested using the hollow cylinder torsional shear device. Typical examples of this kind of data (Yoshimine et al., 1998) are shown in Fig. 3. These results suggest that different residual strengths should be assigned to different parts of the liquefied region depending on the predominant stress conditions.

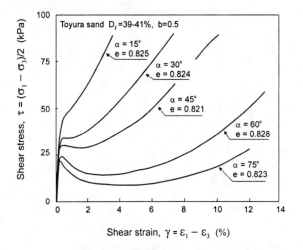

Fig. 3. Effect of stress path on undrained behaviour of Toyoura sand
(Yoshimine et al., 1998)

Finn (1990) drew attention to the important practical implications of the dependence on stress path on the difference between the average steady state strength in the liquefied zones of the San Fernando dam from back-analyses and the values measured in laboratory compression tests :

"The quantitative effect of stress path on steady state strength ---- may account for a substantial part of the difference noticed in the San Fernando studies. This effect is also crucial to a reliable stability analysis"

This use of different shear strengths on potential failure surfaces in limit equilibrium analysis is already part of engineering practice. Bearing capacity under offshore structures in the North Sea is evaluated using compression, simple shear and extension strength data to suit stress conditions at different locations along potential sliding surfaces. This practice seems appropriate also for assessments of post-liquefaction stability. Design decisions based only on compressive undrained tests on loose sands may be potentially unconservative.

Effect of Sample Preparation

Many laboratory studies in liquefaction use samples prepared by moist tamping. However, it frequently results in void ratios that are not accessible to the same sand under deposition conditions in the field. Moist tamped samples are less uniform than pluviated samples. Typical results comparing moist tamped and water pluviated samples are shown in Fig. 4.

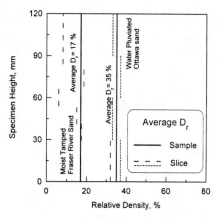

Fig. 4. Uniformity of reconstituted specimens of water pluviated Ottawa sand and moist tamped Fraser River sand (Vaid and Negussey, 1988; Vaid et al., 1999).

Vaid et al. (1988), Yoshimine et al. (1998) Reimer et al. (1997) have demonstrated that the residual strength measured on samples prepared in different ways are quite different (Fig. 5).

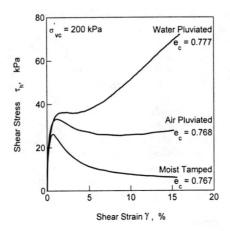

Fig. 5. The effect of sample preparation on undrained simple shear response of
Syncrude sand (Vaid et al., 1999).

Vaid and Sivathayalan (1998) tested frozen samples of two different sands to
determine the residual strength. He then reconstituted the same samples to the same
void ratio using pluviation in water. The reconstituted samples gave residual
strengths very similar to the frozen samples (Fig. 6). In this case, since the same
sand sample was tested in the undisturbed and reconstituted states, there was no
variation in gradation or fines content. This, of course, ensures that the only
difference between the samples was the procedure for creating the sample.

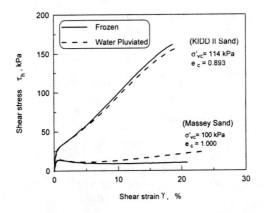

Fig. 6. Comparison of undrained simple shear response of undisturbed in-situ
frozen and equivalent water pluviated sand specimens from
Massey and Kidd Sites (Vaid et al., 1999).

These test results are a strong argument for using pluviation under water to form representative samples of soils that were originally deposited under water or were placed by hydraulic fill construction. The moist tamping method would seem to be more appropriate for unsaturated compacted soils.

Residual Strength as a Function of Effective Confining Pressure

The residual strength was expressed as a fraction of the effective confining pressure in the seismic evaluation of Sardis Dam in 1989 (Finn et al., 1991). Since then the ratio concept has been used in practice on several water-retaining and tailings dams. For the most part the ratio selected has been between 0.06 and 0.1. Similar results have been reported by Baziar and Dobry (1995) and by Ishihara (1993). A value of $S_r/p' = 0.23$ was used for Duncan Dam, based on the testing of frozen samples (Byrne et al., 1994).

Vaid and Thomas (1994), using extension tests, determined the residual strength of Fraser River sand over a range of void ratios and confining pressures. Their results are replotted in Fig. 7; normalized with respect to the effective confining stress. Despite some scatter, the variation of S_r/p' is well represented by a straight line. The ratio varies from about 0.05 at a void ratio of about 0.96, which is the loosest void ratio obtainable by water pluviation, to a value of about 0.2 at a void ratio close to 0.8.

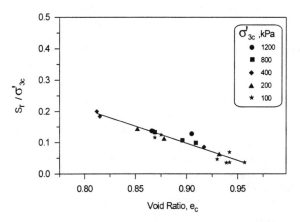

Fig. 7. Relationship between the residual strength normalized by effective confining
pressure and void ratio in extension tests on Fraser River Sand
(Vaid and Thomas, 1994).

Vaid and Sivathayalan (1996) tested Fraser River Sands in simple shear over a range of effective confining pressures and void ratios. The test data again showed the residual strength to be a function of effective confining pressure.

For S_r/p' to be a constant for a given soil, it is necessary for the steady-state line to be parallel to the isotropic consolidation line. This is not generally the case. The ratio is clearly a function of void ratio which is to be expected, since the residual strength is a function of void ratio. However at any given void ratio, considering the scatter in the data, the strength ratio does not appear to be strongly affected by the level of effective confining pressure.

FACTORS CONTROLLING STRENGTH IN THE FIELD

There are a number of factors which affect the assessment of residual strength from case histories which are not reflected in the laboratory tests of uniform samples tested at constant strain rates.

Soil layering can have a major effect on the residual strength that can be mobilized in the field. Seed (1987) drew attention to the possibility of a water film from a liquefied layer collecting under a layer of more impermeable soil. Initially there would be no resistance to sliding in this region. However cracks and drainage would begin quickly. Nevertheless it is plausible to assume that in such a case the average residual strength available during sliding would be lower that what would be inferred from the properties of the liquefied layer alone.

It was shown earlier that the residual strength is a strong function of effective confining pressure. It may not be adequate to develop a correlation between residual strength and $(N_1)_{60}$, if the data is associated with wide ranges in effective overburden pressure. Lo et al. (1991) made an initial attempt to examine the Seed data base from this point of view. They concluded that the effect of overburden was important.

The residual strength derived from the analysis of case histories is also affected by how the analysis is done. The estimation of residual strength is based on equilibrium analysis of sliding on a clearly defined failure surface. Some of the case histories, notably the San Fernando Dam failure, involved large volumes of liquefied material and failure was accompanied by large volumetric distortions, not sliding on distinct failure surfaces. Today it is possible to conduct analyses (large displacement finite element analyses) which can take such distributed shearing into account and hopefully arrive at a more fundamental estimate of the strength at failure.

Castro (NSF, 1997), commenting on the re-analysis of San Fernando Dam, stated that if the accelerations and velocities of the sliding mass had been taken into account, the back-figured steady state strength would be approximately equal to the mean of the statically determined strengths based on the initial and final configurations of the dam (520 psf). This latter stress would allow the dam to fail in a static analysis.

It would be worthwhile to review and augment the Seed data base on case histories. in the light of developments in research and practice since the update by Seed and Harder (1990). The Illinois workshop appointed a committee to re-evaluate the Seed data base on residual strength and to prepare a report. The report is awaited with interest.

POST-LIQUEFACTION DISPLACEMENT ANALYSES

The use of large displacement analysis in evaluating post-liquefaction response and assessing the adequacy of proposed remediation measures has now become part of engineering practice. It was first applied to Sardis Dam in 1989 by Finn et al. (1991). Applications of the method to post-liquefaction deformation analysis and the design of remediation measures have been the subject of previous reviews (Finn, 1993; Finn, 1998). Most assessments of post-liquefaction deformations for embankment dams are now being conducted using large displacement finite element analyses. The analyses are based on an updated Lagrangian formulation and are the basis for computer programs such as TARA-3FL (Finn and Yogendrakumar, 1989) and FLAC (Itasca, 1996). Analysis of this type is essential if post-liquefaction behaviour of an embankment and the location and extent of remediation measures are to be assessed on the basis of deformation criteria. Practice has demonstrated that remediation measures based on displacement criteria are much more cost-effective than those based on the factor of safety approach. Background studies that led to these developments and which show the relationship between the factor of safety approach and the deformation approach are described in Finn (1990).

Validation of Displacement Analysis

There has been little direct validation of displacement analysis based on field data. The few case histories available are poorly defined, both in terms of input motion and in the definition of the deformed shape of the dam. A large number of embankment failures which occurred in Hokkaido during the Kushiro-oki earthquake of 1993 and the Nansei-oki earthquake of 1994, provided a good opportunity for validating post-liquefaction deformation analysis by means of a series of blind tests in which predictions were checked against field data by an independent group.

Between 1995 and 1998, displacement analyses were conducted on many flood protection dikes in Hokkaido, Japan, which had been damaged during the Kushiro-oki and Nansei-oki earthquakes in 1993 and 1994, respectively. The objective was to develop a criterion based on potential post-liquefaction crest settlement to prioritize the remediation of dikes against future earthquakes. A 2-step strategy was adopted for the studies. First, failures of dikes in eastern Hokkaido during the 1993 Kushiro earthquake were to be simulated using the program TARA-3FL (Finn and Yogendrakumar, 1989) If these simulations were satisfactory, then TARA-3FL would be used to predict crest settlements in a number of dikes in western Hokkaido which had significant post-liquefaction displacements during the 1994 Nansei-oki earthquake. All displacement analyses were conducted in Vancouver, Canada, and the predictions were verified by engineers from the Advanced Construction Technology Center (ACTEC), Tokyo, and the Hokkaido Development Bureau on the basis of field data known only to them.

The simulation studies on the Kushiro dikes were considered satisfactory and a major parametric study was approved to investigate the effects of some of the important cross-sectional parameters that control the consequences of liquefaction, such as the thickness of a non-liquefiable layer overlying the liquefied layer, the thickness of the liquefied layer itself, and the height and side slopes of the dikes. The effects of these parameters were characterized by the settlements of the crests of the dikes after liquefaction.

Estimation of Crest Settlements

The crest settlements were estimated first for the dikes with side slopes 1:2.5 as shown in Fig.8.

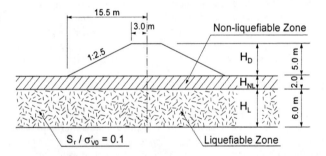

Fig. 8. Typical cross-section of dike for analysis.

The thicknesses of the liquefied and nonliquefied layers were varied and the resulting displacements are plotted in nondimensional form in Fig.9.

The nondimensional computed crest settlements, S/H_D are shown by the curve in Fig. 9. The equation of the curve is given by

$$\frac{S}{H_D} = 0.01 \exp\left(0.922 \frac{H_D}{H_{NL}} \frac{H_L}{H_{NL}}\right) \qquad (1)$$

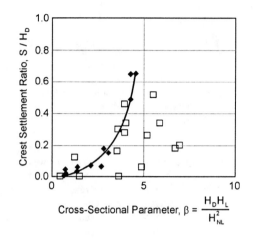

Fig. 9. Comparison of computed dike settlements with settlements
during the 1994 earthquake.

where S is the crest settlement, H_D, is the height of the dike; H_L and H_{NL} are the thicknesses of the liquefiable and non-liquefiable layers, respectively. The residual strength ratio was $S_r/\sigma'_{vo} = 0.10$. This curve was adopted for predicting crest settlement in other dikes which were not analysed.

Engineers from ACTEC and the Hokkaido Development Bureau compared the measured crest settlements from a wide variety of dikes in western Hokkaido which underwent noticeable displacements during the Nansei-oki earthquake in 1994 with those predicted by Eqn. 1. The data points and the prediction curve are plotted in Fig. 9. The black points represent the settlements from the idealized analyses done to develop the curve; the open points represent dikes which had not been analysed. The agreement was very good for dikes with slopes of 1:2.5, but the field data showed that the side slopes had an important affect on the crest settlement and that separate criteria would be necessary for two other predominant slopes; uniform side slopes of 1:5, and unequal slopes, 1:5 and 1:10. These are represented by points lying substantially to the right of the curve. It can be seen that the criterion based on crest settlement can be useful for deciding which dikes should be remediated first. The larger the predicted settlement, the more urgent the need for remediation. The Hokkaido dikes study is an important case history because it is the only instance in which post-liquefaction displacement analysis has been validated independently in blind tests in a large number of earth structures undergoing different levels of post-liquefaction displacements.

CONCLUSIONS

Large displacement finite element analysis for estimating post-liquefaction displacements and assessing the effectiveness of proposed remediation measures is being used increasingly in practice. Since 1989 this type of analysis has been used in over 15 major embankment dams, both water retaining dams and tailings dams. Experience indicates that using displacement rather than factor of safety criteria for assessing the amount of remediation required leads to reduced costs. Therefore use of this type of analysis is likely to grow.

The residual strength is a key parameter in any type of analysis of post-liquefaction stability and has a major impact on remediation costs. However the estimation of residual strength for use in engineering practice is in an unsatisfactory state. Despite the findings from years of research by reputable groups based on high quality laboratory studies, there are still very divergent opinions on the basic factors controlling residual strength. A framework of understanding with broad support could be achieved with the present state of knowledge outlined above given the right leadership and under the right auspices. This field of endeavor needs a committee to do for it what the committee chaired by Professor Whitman did for the triggering of liquefaction in 1985 (NRC, 1985) and the committee chaired by Professors Youd and Idriss did in 1997 (NCEER, 1997).

The Geo-Institute has the prestige and capability to set up such a committee, develop relevant terms of reference and to help secure funding to support the work of the committee. The potential impact of bringing some order and coherence to this field is so important from both a safety and economics point of view that a very strong case can be made for support. The ultimate objective of such an undertaking should be to provide a coherent framework of understanding for practicing engineers.

REFERENCES

Baziar, M.H. and Dobry, R. (1995). "Residual Strength and Large-Deformation Potential of Loose Silty Sands", ASCE Journal of Geotechnical Eng. Div., Vol. 121, No. 12, pp. 896-906.

Byrne, P.M., Imrie, A.S. and Morgenstern, N.R. (1994). "Results and Implications of Seismic Performance Studies for Duncan Dan", Canadian Geotechnical Journal, Vol. 31, No. 6, December, pp. 979-988.

Castro, G. (1969). "Liquefaction of Sands", Harvard Soil Mechanics Series No. 81, Pierce Hall, Cambridge, Massachusetts.

Castro, G. (1997). "Post-Liquefaction Shear Strength from Case Histories", Preliminary Proceedings of the Workshop "Post-Liquefaction Shear Strength of Granular Soil" (Editors T.D Stark, S.L. Kramer and T.L. Youd), National Science Foundation, Washington, DC., pp. 61-67.

Dobry, R. (1995). "Liquefaction and Deformation of Soils and Foundations Under Seismic Conditions", Proc., 3rd International Conference, on Recent Advances in Geotechnical Earthquake Engineering and Soil Dynamics, Vol. 3, April, St. Louis, Missouri, pp. 1465-1490.

Emery, J.J., Finn, W.D. Liam and Lee, K.W. (1973). "Uniformity of Saturated Sand Specimens in Evaluation of Relative Density and its Role in Geotechnical Projects Involving Cohesionless Soils", ASTM, STP 523, American Society for Testing and Materials, pp. 182-194.

Finn, W.D. Liam and Yogendrakumar, M. (1989). "TARA-3FL: A Program for Analysis of Flow Deformations in Soil Structures with Liquefied Zones", Soil Dynamics Group, Department of Civil Engineering, University of British Columbia, Vancouver, B.C.,

Finn, W.D. Liam, (1990). "Analysis of Post-Liquefaction Deformations in Soil Structures", Invited Paper, Proc., H. Bolton Seed Memorial Symposium, University of California, Berkeley, Editor J.M. Duncan, Bi-Tech Publishers, Vancouver, Canada, Vol. 2, May 9-11, pp. 291-311.

Finn, W.D. Liam, Ledbetter, R.H., Fleming, R.L. Jr., Templeton, A.E., Forrest, T.W. and Stacy, S.T. (1991). "Dam on Liquefiable Foundation: Safety Assessment and Remediation", Proc., International Workshop on Remedial Treatment of Potentially Liquefiable Soils, Tsukuba Science City, Japan, January, 37 pages.

Finn, W.D. Liam. (1993). "Seismic Safety Evaluation of Embankment Dams", Proc., International Workshop on Dam Safety Evaluation, Vol. 4, Grindewald, Switzerland, 26-28 April, pp. 91-135.

Finn, W.D. Liam (1998). "Seismic Safety of Embankment Dams Developments in Research and Practice 1988-1998", Proc., Geotechnical Earthquake Engineering in Soil Dynamics III, Edited by P. Dakoulas, M. Yegian and R.D. Holtz, Geotechnical Special Publication No. 75, ASCE, Vol. 2, pp. 812-853.

Itasca (1996). FLAC (Version 3.3) (1996). Itasca Consulting Group Inc., 708 South 3rd Street, Suite 310, Minneapolis, Minnesota 55415.

Ishihara, K. (1993). "Liquefaction and Flow Failure During Earthquakes", Geotechnique, Vol. 43, No. 3, pp. 351-415.

Lade, P.V. and Yamamuro, J.A. (1998), "Physics and Mechanics of Soil Liquefaction", Proc., International Workshop, Johns Hopkins University, Baltimore, Maryland, A.A. Balkema, Rotterdam/Brookfield, 1999.

Lo, R.C., Klohn, E. and Finn, W.D. Liam (1991), "Shear Strength of Cohesionless Materials Under Seismic Loadings", Proceedings, IX Pan-American Conference on Soil Mechanics and Foundation Engineering, Volume III, Chile, August 1991, Chile, pp. 1047-1061.

National Research Council (1985). "Liquefaction of Soils During Earthquakes", Report of Committee on Earthquake Engineering, Washington, DC.

National Science Foundation (NSF) (1997). "Post liquefaction Shear Strength of Granular Soils", Proc., Workshop on Post-Liquefaction Shear Strength of Granular Soils, Edited by T.D. Stark, S.L. Kramer, and T.L. Youd. April 18-19.

NCEER (1997). Proc., Workshop on Evaluation of Liquefaction Resistance of Soils, Technical Report No. NCCER-97-0022. Edited by T.L. Youd and I.M. Idriss.

National Center for Earthquake Engineering Research, University of Buffalo, Buffalo, New York.

Poulos, S.J., Castro, G. and France, W. (1985). "Liquefaction Evaluation Procedure", Journal of the Geotechnical Engineering Division, ASCE, Vol. 111, No. 6, pp. 772-792.

Riemer, M.F. and Seed, R.B. (1997)." Factors Affecting Apparent Position of Steady State Line", ASCE Journal of Geotechnical and Geoenvironmental Engineering, Vol. 123, No. 3, pp. 281-288.

Seed, R.B. and Harder, Jr. L.F. (1990). "SPT-Based Analysis of Cyclic Pore Pressure Generation and Undrained Residual Strength", Proc., H. Bolton Seed Memorial Symposium, University of California, Berkeley, Editor J.M. Duncan, Bi-Tech Publishers, Vancouver, Canada, Vol. 2, May 9-11, pp. 351-376.

Seed, H.B. (1987). "Design Problems in Soil Liquefaction", Journal of Geotechnical Engineering, ASCE, Vol. 113, No.7, pp. 827-845, August.

Seed, H.B., Seed, R.B., Harder, Jr., L.F., Jong, H.-L. (1989). "Re-evaluation of the Lower San Fernando Dam", Report No. 2, Examination of the Post-Earthquake Slide of February 9, 1971, Dept. of the Army, US Army Corps of Engineers, Washington, DC 20314-1000.

Utahayakumar, M. and Vaid, Y.P. (1998). "Liquefaction of Sands Under Multi-axial Stresses", Canadian Geotechnical Journal, Vol. 35, No.2, pp. 273-283.

Vaid, Y.P. and Chern, J.C., "Cyclic and Monotonic Undrained Response of Saturated Sands", ASCE National Convention, Session - Advances in the Art of Testing Soils Under Cyclic Loading, Detroit, pp. 120-147, October 21-25, 1985.

Vaid, Y.P. and Negussey, D. (1988). "Preparation of Reconstituted Sand Specimens", Advanced Triaxial Testing of Soils and Rock, ASTM STP 977, pp. 405-417.

Vaid, Y. P., and Sivathayalan, S. (1996). "Static and Cyclic Liquefaction Potential of Fraser Delta Sand in Simple Shear and Triaxial Tests", Canadian Geotechnical Journal, Vol. 33, No. 2, pp. 281-289.

Vaid, Y.P. and Sivathayalan, S. (1998). "Fundamental Factors Affecting the Liquefaction Susceptibility of Sands", in Physics and Mechanics of Liquefaction, Balkema, Rotterdam/Brookfield,1999, pp. 105-120.

Vaid, Y. P., Sivathayalan, S., and Stedman, D. (1999). "Influence of Specimen Reconstituting Method on the Undrained Response of Sand", ASTM Geotechnical Testing Journal, Vol. 22, No. 3, pp. 187-195.

Vaid, Y.P. and Thomas, J. (1994). "Post-Liquefaction Behaviour of Sand", Proc., 13th International Conference on Soil Mechanics and Foundation Engineering, New Delhi, Vol. 3, pp. 1305-1310.

Yoshimine, M., Ishihara, K., and Vargas, W. (1998). "Effect of Principal Stress Direction and Intermediate Principal Stress on Undrained Shear Behaviour of Sand", Soils and Foundations, to appear in 1998.

DYNAMIC DEFORMATION ANALYSIS AND THREE-DIMENSIONAL POST-EARTHQUAKE STABILITY ANALYSIS FOR CASITAS DAM, CALIFORNIA

John W. France,[1] Member, ASCE, Tiffany Adams,[2] Member, ASCE, John Wilson,[3] Member, ASCE and Dave Gillette,[4] Member, ASCE

ABSTRACT

Casitas Dam is a 335-foot high, zoned earthfill embankment located in a highly seismic area, near Ventura, California. The U.S. Bureau of Reclamation (Reclamation) concluded that the foundations of the dam were susceptible to liquefaction in the event of an earthquake, and that liquefaction of the foundations could lead to failure of the embankment and uncontrolled release of the reservoir. Reclamation developed designs for dam safety modifications to increase the seismic stability of the downstream section of the dam, but the planned modifications do not include treatment of the upstream foundations. Two-dimensional dynamic deformation analyses and three-dimensional post-earthquake stability analyses were performed to evaluate the expected performance of the untreated upstream slope during and after significant earthquake shaking. The deformation analyses indicated that significant deformations would be expected in the upstream section of the embankment dam, but that freeboard would be maintained at the downstream edge of the crest of the modified dam. The three dimensional stability analyses indicated that three-dimensional factors of safety for the upstream slope that were between 10% and 35% higher than the corresponding two dimensional factors of safety. Based on these results, the two-dimensional stability analyses may be over-estimating the deformations in the upstream section of the dam. This paper includes a discussion of the methods used in the analyses, the results obtained, and limitations and other issues that had to be addressed to complete the analyses.

INTRODUCTION

The U.S. Bureau of Reclamation (Reclamation) has recently completed modification designs to alleviate potential seismic dam safety concerns for Casitas

[1] P.E., Vice President, URS Greiner Woodward Clyde, 4582 S Ulster St., Denver, CO 80237
[2] Geotechnical Engineer, URS Greiner Woodward Clyde, 4582 S Ulster St., Denver, CO 80237
[3] P.E., Civil Engineer, U.S. Bureau of Reclamation, Denver Federal Center, Building 67, Denver, CO 80225-0007
[4] P.E. Civil Engineer/Technical Specialist, U.S. Bureau of Reclamation, Denver Federal Center, Building 67, Denver, CO 80225-0007

Dam. Analyses completed by Reclamation had identified the existence of potentially liquefiable soils in the foundation beneath the dam. Two-dimensional dynamic deformation analyses and three-dimensional post-earthquake stability analyses were completed as part of the engineering effort to support the modification designs.

Casitas Dam, the main feature of the Ventura River Project, was completed in 1959, and is located on Coyote Creek about 6.5 miles northwest of Ventura, California. The dam is a zoned earthfill embankment, with a structural height of 335 feet and a capacity of 250,835 acre-feet at the top of active conservation, elevation 567 feet. The crest of the embankment is 2,000 feet long and 40 feet wide at elevation 585 feet. The foundation of the dam consists of alluvial deposits with a thickness of 40 to 45 feet, overlying bedrock. The alluvial deposits are composed of an upper sandy layer and a lower gravelly layer. The dam is in a highly active seismic area, and recent studies have identified several potential seismic sources. The controlling source is the Red Mountain Fault, which is capable of producing a maximum credible earthquake (MCE) of magnitude (M_w) 6.9, at a distance of 3 kilometers from the site. Failure of the dam could result in severe economic damage and loss of many lives in communities downstream.

In situ tests, including standard penetration tests (SPT), Becker hammer tests, and crosshole shear wave velocity measurements, were completed to evaluate the liquefaction potential of the alluvial foundation soils. Based on the test results, it was concluded that sections of the upper alluvium beneath the upstream and downstream shells of the embankment, are likely to liquefy in the event of a large earthquake. It was also concluded that the lower alluvium would likely not liquefy, but may be susceptible to some increase in pore water pressure and reduction in strength. Based on the SPT test results, a conservative estimate of the liquefied (post-earthquake) shear strength of the upper alluvium is about 600 psf. For this estimated strength, two-dimensional post-earthquake stability analyses indicate factors of safety significantly below 1.0 for the downstream slope and between 1.0 and 1.2 for the upstream slope. Based on these analysis, it was concluded failure of the embankment would be likely during the design earthquake loading and that dam safety modifications were required.

The modifications that have been designed by Reclamation include excavation and replacement of foundation soils at the downstream toe and construction of a downstream buttress and filters. No modifications are planned for the upstream section of the dam. Reclamation contracted with URS Greiner Woodward Clyde (URSGWC) to complete two-dimensional dynamic deformation analyses and three-dimensional post-earthquake stability analyses for the planned modifications. Those analyses are described in this paper.

DYNAMIC DEFORMATION ANALYSIS

Methodology and Analysis of Existing Conditions

The dynamic deformation analyses were completed using the finite difference computer program FLAC (Fast Lagrangian Analysis of Continua, Itasca, 1995). For Casitas Dam (shown in plan on Figure 1), the existing cross-section shown in Figure 2a was modeled by the finite element mesh shown in Figure 2b. The soil properties used for the analyses are summarized in Table 1. Properties were selected based on the results of field investigations, laboratory tests, and published correlations. It should be noted that the cross-section shown in Figure 2a does not exist in the field as a continuous section. Because of the S-shaped configuration of the river valley (in plan view – see Figure 1), the cross-section shown in Figure 2a is actually a composite of upstream and downstream sections through the respective maximum-height sections of the embankment.

Two input earthquake records were considered in the analyses: a record from the Sylmar Hospital in the 1994 Northridge earthquake and a record from the Los Gatos Presentation Center in the 1989 Loma Prieta earthquake. The positive and negative polarities were considered for both records, resulting in consideration of four input records. All four records were considered and compared in initial analyses. Calculated deformations were similar for all four records, and the Sylimar north-south positive polarity record was selected for all subsequent analyses, because it resulted in slightly higher calculated deformations than the other three records. In all cases, potentially liquefied layers were assigned non-liquefied strengths at the beginning of the analysis, and then the strengths for these layers were instantaneously reduced to the liquefied strengths at 3 seconds into the Sylmar record and at 5.5 seconds into the Loma Prieta record. This approach was selected because there was little strong motion prior to these times in the two records and significant strong motion very shortly afterward.

Initial validation of the deformation model was completed by running analyses for the existing conditions, to determine if these analyses would predict failure of the structure. The deformed mesh resulting from dynamic analyses for existing conditions with a liquefied strength of 600 psf is presented in Figure 3. For this case, analysis computations were terminated after 10.5 seconds, because of excessive deformations of some elements of the model. Deformations in these elements were so great that the computations became unstable. The mesh was still deforming significantly at 10.5 seconds, and much greater computed deformations would have resulted if the analysis could have been continued. However, this analysis was not continued, because the results at 10.5 seconds already indicated that the embankment crest had moved downward to elevation 557 feet, 10 feet below the reservoir level and 28 feet below the initial crest elevation. Consequently, it was concluded that the failure indicated by post-earthquake stability analyses was confirmed by the deformation analysis.

Initial Modification Cross-Section

Next, dynamic analyses were completed for an initial embankment modification cross-section developed by Reclamation. The initial modification cross-section is shown in Figure 4a. The initial modification cross section included the following features: 1) excavation and replacement of foundation soils at the downstream toe, 2) addition of a downstream stability berm, and 3) a 30-foot widening of the existing crest and downstream slope above the stability berm. The resulting crest width is 70 feet, compared to an initial crest width of 40 feet. The downstream excavation would extend fully through the upper alluvium layer, which was judged to be liquefiable. The stability berm and the 30-foot wide overlay on the downstream slope above the berm would include a filter layer to control leakage through transverse cracks caused by the earthquake. The modifications would not include any changes or treatments to the upstream slope or the upstream foundations.

The FLAC mesh was modified to reflect the addition of the berm and the downstream overlay and the replacement of the downstream foundation materials (Figure 4b). The upper alluvium was still modeled as a liquefiable material at locations where it was not replaced. To evaluate the sensitivity of the results to variations in liquefied strength, dynamic analyses were completed for three values of liquefied strength for the upper alluvium: 400 psf, 600 psf, and 800 psf. In all three cases, gravity loads were maintained after the cessation of earthquake shaking, to evaluate the post-earthquake performance of the structure. The results indicated that deformations of the downstream section of the embankment stopped very shortly after the end of earthquake motions, for all three cases. For the 800 psf case, the analyses indicated that deformations of the crest and the upstream section of the embankment also stopped, and that the structure was stable very shortly after the earthquake motion ended. However, for the cases of 400 psf and 600 psf, deformations of the crest and the upstream section of the embankment continued after the earthquake motion had stopped. In both cases the analysis computations were terminated because excessive deformations of some elements of the mesh resulted in computational instability. In both cases, the analyses indicated that significant deformations of the crest and the upstream section of the embankment were still occurring when the analyses were terminated. Deformed meshes at the end of dynamic analysis for all three cases are shown in Figure 5.

The post-earthquake deformation analyses for the 400 psf and 600 psf cases were continued through a remeshing process. This process consisted of reconfiguring the mesh elements that had experienced excessive deformation, and then continuing the gravity-load analysis. The remeshing process was repeated until the analysis indicated stability (i.e. no further deformations under gravity load). The fully liquefied strength was maintained for the remeshing analyses; that is, no allowance was made for possible recovery of some of the strength lost during liquefaction. The 400 psf case required three remeshings and the 600 psf case required two remeshings. In both cases, the deformations were very small when gravity loads were applied after the last remeshing; in fact, for the last application of gravity load in each case, the undeformed and deformed meshes were virtually indistinguishable. Therefore, the analyses had practically reached stability before the final remeshing in both cases. The deformed meshes after stability was reached are shown for both cases in Figure 6.

Enlarged views of the crest of the deformed embankment are shown in Figure 7 for all three strength cases. All three results shown in Figure 7 are for the final deformed meshes after stability was reached, including the results of the remeshing analyses for the 400 psf and 600 psf cases. Based on these results, Reclamation concluded that additional modifications to the design were required.

Second Modification Cross-Section

Reclamation developed a second modification cross section, as shown in Figure 8a. The additional modifications included in this cross section are: 1) an additional 20-foot increase in the width of the downstream slope overlay and the crest, to a total width of 90 feet, and 2) inclusion of geogrid reinforcement in the downstream slope overlay for the upper 30 feet. The FLAC mesh was revised to reflect both of these changes (Figure 8b). The geogrid reinforcement was modeled in the FLAC mesh by the inclusion of tensile elements, with strength and deformation characteristics based on the properties of commercially available geogrids. Deformation analyses were completed for the second modification cross section only for the case of a liquefied strength of 400 psf. As for the analyses for the first modification section, the analyses were extended through three remeshing analyses, until stability was indicated, and the resulting deformed mesh is shown in Figure 9. The deformed meshes in the crest area for the 400 psf cases are compared in Figure 10, for the first and second modification cross sections.

Summary of Deformation Analysis Results

The horizontal and vertical deformations of the upstream and downstream edges of the crest are summarized in Table 2 for the cases discussed above. That table also includes the estimated remaining freeboard heights with respect to the highest remaining points on the crest. Figure 11 presents a graphical comparison of horizontal and vertical deformations of the crest for the initial and second modification cross sections, both for a liquefied strength of 400 psf. In that figure, the deformations are plotted versus distance from the upstream edge of the crest. From Figure 11, it is seen that the improvement provided by the second modification cross section results from two things: (1) the geogrids holding the downstream section of the crest together as a block and (2) the wider downstream crest resulting in less settlement at the downstream edge of the crest. The figure also illustrates that the deformations of the upstream section of the dam are practically the same for both cases.

The results of the analyses are compared another way in Table 3. This table presents a summary of the "freeboard volumes" in the embankment before and after the deformation analyses. "Freeboard volume" is defined as the volume of the embankment above the reservoir water level (elevation 567 feet). For comparison purposes, the freeboard area for the existing embankment is 1,530 ft^3/ft, as noted in the table.

Based on these results, Reclamation concluded that the second modification cross section would provide acceptable protection against catastrophic failure of the dam and release of the reservoir as a result of the design earthquake. The upstream section of the dam will likely experience significant deformations and repairs would be required after the earthquake. The magnitude of the deformations will depend significantly on the liquefied strength of the foundations, and the deformations may be relatively small, if the liquefied strengths are as little as 25% higher than the current best estimate. In addition, deformations may be less than predicted by the two-dimensional deformation analyses, if the post-earthquake stability factors of safety are higher because of three-dimensional effects, as discussed below. As noted previously, the design of the berm and downstream overlay includes filter zones to protect against a piping failure resulting from seepage through transverse cracks caused by deformations of the upstream section of the embankment.

The dynamic deformation analyses included several other analysis cases beyond those discussed above. However, those cases could not be covered because of the length limitations for this paper. For example, analyses were completed to evaluate the effects of reduced strength in the lower alluvium, and it was found that the results were not significantly affected for reasonable estimates of strength reduction for this material.

THREE-DIMENSIONAL STABILITY ANALYSIS

Methodology

Three-dimensional slope stability analyses were performed using the computer program CLARA (O. Hungr, Geotechnical Research Inc., 1988). CLARA is based on an extension of the Bishop Simplified Method of slices to three dimensions and to non-circular cases. In the CLARA program the sliding mass is divided into evenly-spaced, three-dimensional columns. As is the case for the two-dimensional Bishop's Simplified method, both vertical force and moment equilibrium are satisfied and vertical inter-column forces are neglected.

The current version of the CLARA program has some limitations. One of the major limitations is that the size of the three-dimensional columns used in the CLARA model is limited because of the computer memory accessible to the program. The column size used in the Casitas model was 25 feet in the downslope direction and 20 to 25 feet in the transverse direction. It would have been desirable to use a smaller column size to increase the accuracy of the computations, but this was not possible because of the memory limitation. Another limitation is that the CLARA program does not include a routine to search for the surface with the minimum factor of safety. Therefore "approximate lowest factors of safety" were estimated using iteration and judgment. Both truncated-ellipsoidal-shaped surfaces and wedge-shaped surfaces were considered in the analyses. However, because of the limitation on column size and the geometry of the three dimensional surfaces, the results obtained with the wedge-shaped surfaces were erratic and were judged to be unreliable. Therefore all of the three-dimensional results presented in this paper are for truncated-ellipsoidal-shaped surfaces.

Analysis Sections and Material Characterization

For the completion of this work, nine study cross-sections were developed (Sections A through G, Y, Z), centered in the area where the Coyote Creek channel passes beneath the upstream slope of the dam, approximately between stations 11+50 and 20+00. The locations of these sections are shown on Figure 1.

A representative two-dimensional stability cross-section, Section C, is shown in Figure 12, along with a table of the material properties for the cross-section. For the three-dimensional slope stability analyses, a liquefied shear strength of 600 psf was used for the upper alluvial material. The analysis section in this study is an upstream slope, much of which is below the reservoir level. A standard approach to evaluate the stability of the upstream slope of a dam is to use total weights and boundary water forces. However, the CLARA program is not capable of accepting boundary pressures from water. Therefore, an approach using buoyant unit weights for the materials below the phreatic surface was adopted. Seepage forces were not included in the analysis because there is no appropriate way of incorporating them in the CLARA program. This omission was considered to be acceptable because the seepage forces associated with the mostly horizontal phreatic surface would be relatively small.

Analysis Cases and Results

CLARA analyses were performed for 3 cases: 1) the three-dimensional surface with approximately the lowest overall factor of safety, 2) the three-dimensional surface with approximately the lowest factor of safety that intercepts the crest at a point 40 feet upstream of the downstream edge of the crest, and 3) the three-dimensional surface with approximately the lowest factor of safety that passes through the downstream slope 5 feet (vertically) below the crest.

Standard CLARA Analysis - The first group of analyses performed considered the three cases listed above and used a standard Mohr-Coulomb strength characterization in the CLARA program. These analyses do not consider the contribution of horizontal earth pressures to the sliding resistance generated along the sidewalls. Two-dimensional slope stability analyses were also conducted using the CLARA program (using the Bishop Simplified method) and UTEXAS3 (using Spencer's method), for the three cases listed above for comparison purposes.

The results for case 2 are shown in Figure 13. The three-dimensional slide surface is shown in orthographic relief and in plan. The analyses were performed using a symmetric mesh in order to get the most accurate results with the memory limitations of the CLARA program, and as such, only half of the slide surface is shown. The three-dimensional failure surfaces identified for cases 1 and 3 were similar in shape and their outlines are shown along with the outline of the case 2 failure surface (top of scarp, base of scarp, sidewalls, and toe) in Figure 14. The base of scarp and toe lines are essentially the same for all three cases, but the downstream extent of the slide (top of scarp line) is not. The critical three-dimensional surface for each case was limited to a width of 750 feet, considered to be the widest that the subsurface topography would allow, and therefore should represent the lowest three-dimensional effects. The two-dimensional factors of safety computed at the maximum section of the three-dimensional failure surfaces are shown on Figure 15.

The results show that the effect of consideration of the three-dimensional conditions on the computed factor of safety is an increase of about 11 to 12 percent when compared to the two-dimensional analysis results. The two-dimensional factor of safety computed with Bishop's method and buoyant unit weights in the CLARA program and Spencer's method and total unit weights with boundary water forces in the UTEXAS3 program were similar, indicating that the use of Bishop's Simplified method and buoyant unit weights for the two-dimensional, and, therefore, three-dimensional slope stability analysis is reasonable.

Sensitivity of the CLARA Analysis Results - The results of three-dimensional slope stability analysis using the CLARA program indicate that there is a moderate gain in computed factor of safety when compared to two-dimensional results. However there are several factors which would affect the analysis results, such as: 1) the uncertainties of the extent of the liquefiable upper alluvium layer under the upstream slope of the dam, 2) the shape of the sliding surface within the liquefiable layer, and 3) the incorporation of sliding resistance on the sidewalls of the slide.

The increase in computed three-dimensional factor of safety when compared to the two-dimensional results is affected by the width of slide considered. The three cases (Case 1, Case 2, and Case 3) presented on Figures 14 and 15 were used to perform a parametric analysis in which the widths of the three-dimensional analysis sections were varied. The three-dimensional factor of safety decreased with increasing slide width tending towards the two-dimensional analysis results at large slide widths, as would be expected. The results presented in this report are for the estimated maximum failure scarp width that could develop, considering the geometric constraints at Casitas Dam. If the subsurface conditions are different than those estimated for this model, and the extent of the liquefiable layer near the upstream toe of the dam is wider or narrower, the computed three-dimensional factor of safety for all three cases would be affected.

The computed three-dimensional factor of safety is also affected by the percentage of the potential failure surface that passes through the liquefied upper alluvium layer. For the ellipsoidal potential failure surfaces presented in this report, the shape of the curve representing the intersection of ellipsoidal surface with the horizontal plane within the liquefied layer is parabolic: narrow near the dam crest and widening towards the upstream toe. It is possible that a more critical surface with a lower factor of safety would be one that maximized the sliding surface in the liquefied layer by remaining wide throughout the length of the slide, possibly similar to a wedge-shaped failure surface. Unfortunately, the limitations of the program prevent appropriate analysis of wedge-shaped failure surfaces. Changes in kinematic constraints and possible increases in shear resistance from the side surfaces could possibly counteract this effect of a larger area in the lower shear strength liquefied layer.

The three-dimensional factor of safety computed by the CLARA program does not consider the full resistance to sliding provided by the non-horizontal side areas of the potential failure surface. The three-dimensional stability method used by the CLARA program is an extension of the two-dimensional Bishop's Simplified method. As with most stability methods, it calculates frictional resistance on the potential failure surface based on the overburden pressure, usually at the midpoint of the vertical slice or column. However, the frictional resistance generated in the field along the non-horizontal side areas of the three-dimensional potential failure surface would be higher because of the effect of horizontal earth pressures present within the slope.

The conventional two-dimensional slope stability analysis method does not incorporate the contribution of the effective horizontal stresses in estimating the shear resistance on the potential failure surface along the "sloping back portion" of the section, such as the backscarp of any one of the two-dimensional failure surfaces shown in Figure 15. Considering the kinematics of the section of the failure mass represented by Figure 15, it is reasonable and appropriate not to incorporate the effects of horizontal stresses along the sloping back portion of a three-dimensional surface. However, the situation is different for the kinematics of side surfaces. Along side surfaces, there is no conventional way of evaluating shear resistance because these portions of potential failure surface are not included in two-dimensional slope stability analysis.

In this regard, previous studies (e.g., Wright and Duncan, 1973) indicated that normal stresses on the potential failure plane, calculated using finite element methods, can be used to estimate the potential shear resistance. This approach has been used by others, for example, in a recent study by Stark and Eid (1998), to adjust the shear strength assigned to areas of three-dimensional potential failure surfaces.

Refined **CLARA Analysis** - A second suite of analyses completed for this study included a refinement in the method by which shear strength was assigned in the three-dimensional slope stability analyses using the CLARA program. The refinement was intended to reflect various postulated potential shear resistance values along the sides of the potential failure surface.

Refinement of Shear Resistance Estimates - Where appropriate, shear resistance values were estimated using the approach discussed below, which used in part the results of vertical and horizontal effective stresses calculated in the deformation analysis study discussed in the first half of this paper.

The stress state at the bottom of each column of the potential failure surface in the three-dimensional CLARA model was estimated based on the results of two-dimensional FLAC analysis at the corresponding location. This stress state was then used to calculate an equivalent shear resistance along the bottom of the column in the direction of potential movement using the following equation:

$$C_i = (\sigma_v' . \cos \alpha + f . \sigma_h' . \sin \alpha) \tan \phi$$

where:

C_i = equivalent shear resistance at the bottom of column "i"

σ_v', σ_h' = vertical and horizontal effective stress at the bottom of the column,

α = angle between the horizontal and the plane tangent to the bottom of the column,

f = contribution factor for the horizontal earth pressure

   = 0 for standard CLARA slope stability analysis

   $\leq$ 1 used in this approach, and

ϕ = friction angle of the material along the bottom of the column.

The equivalent cohesion along the potential failure surface was then evaluated as:

$$C = \Sigma(C_i.A_i) / \Sigma A_i$$

where:

C = equivalent cohesion along the potential failure surface
A_i = area of the bottom of column "i"

This equivalent cohesion was then used in the CLARA program as a specified cohesion with zero friction angle to compute the shear resistance along the sides and backslope of the potential failure surface.

Analysis Cases - Using the approach described above, four scenarios were evaluated for each of the three cases listed above. The four scenarios are schematically illustrated in Figure 16. In this figure, the potential failure surface is shown in plan view as a crescent-shaped closed curve. The left U-shaped curve corresponds to the intersection of the sloping portion of the potential failure surface with the bottom horizontal plane (base of scarp). Similarly, the right U-shaped curve corresponds to the intersection of the sloping portion of the potential failure surface with the ground surface (top of scarp). The contribution factor of the horizontal earth pressure, f, is shown as a function of location on the failure surface for the four scenarios.

It is important to note that for scenarios A,B,C, and D, an equivalent cohesion value was assigned only to the inclined portion of the failure surface (the back and side slopes of the sliding mass). The resistance along the bottom horizontal plane (in the upper alluvium layer) was not modified in this process.

In scenario A, the value of f was set to zero throughout the inside of the crescent shape (the sloping portion of the potential failure surface). In scenario B, the value of f was set to one throughout the inside of the crescent shape. In scenario C, the value of f was set to one along the sidewalls of the potential failure surface and to zero along the backwall. In scenario D, the value of f was linearly varied along the sidewall of the potential failure surface from zero along the backwall to full effect (100%) at the toe of the potential failure surface.

In our opinion, Scenarios C and D provide reasonable bounds to the effects of horizontal stresses on the three-dimensional safety factors for the upstream slope. Scenario A is essentially the same as the standard CLARA analysis (with no consideration of horizontal stresses), and Scenario B is not judged to be a reasonable reflection of three-dimensional factors of safety because it overestimates the effects of horizontal stresses by including the effect of horizontal earth pressures along the backwall, where the kinematics of movement are such that they would be expected to significantly decrease horizontal stresses. Scenario B is included only to provide an upper bound to the effect of horizontal stresses on the three-dimensional safety factors for the upstream slope.

Analysis Results - The results of the refined CLARA analyses are summarized in Table 4. Results for both two-dimensional and three-dimensional models are presented for the surfaces considered in the standard CLARA analysis Cases 1, 2, and 3. The results for the standard CLARA analyses for Cases 1, 2, and 3 are also included in Table 4, in the "3-D Standard, Coarse Mesh" column.

The results from the three-dimensional refined CLARA analyses using the same column spacing as used in the standard CLARA analyses are presented in the column in Table 4 labeled "3-D Refined, Coarse Mesh." Factors of safety reported in the table for the refined analysis correspond to an average of the calculated values for Scenarios C and D, considered to be the more appropriate scenarios for incorporating the effects of horizontal stresses on the sides of the sliding surface. The table also includes a columns labeled "% Increase in 2-D FS for Refined vs. Coarse Mesh." As discussed previously in this report, the column size that could be used in the CLARA model was limited by the computer memory that the program could access. Two-dimensional coarse mesh analyses completed using the same column spacing that was used in the three-dimensional model were compared with two-dimensional fine mesh analyses completed using the smallest column size allowed for two-dimensional analyses within the memory limitation of the CLARA program. The two-dimensional fine mesh analyses used one foot column widths, and are the same results as those discussed earlier in the paper and on Figure 15.

The results summarized in Table 4 show that the two-dimensional fine mesh model results in significantly higher factors of safety than the two-dimensional coarse mesh model. The fine mesh model results should be more accurate than the coarse mesh model, because the fine mesh model more closely approximates the slice widths that are used in conventional two-dimensional stability analysis programs. The increase in factor of safety for the fine mesh model can be attributed to a better discretization of the columns in the area where the failure plane crosses the boundary between the back scarp (in frictional material) and the base of the slide (in liquefied material). In CLARA, the shear strength assigned to a soil column is based on the conditions at the coordinates of its center along the base surface. In the coarse mesh model, the columns near the bottom material boundaries usually are assigned to the lower strength material (liquefied residual strength), even though significant portions of the bases at the columns are in the frictional material; whereas for the finer mesh model the columns are smaller and assigning strengths based on conditions at their centers does not introduce as much error. The inaccuracy of the coarse mesh results also depends somewhat on the shape of the failure plane. The difference between coarse and fine mesh results is greater when the transition between the back scarp and the base of the slide is more abrupt (Cases 1 and 2). When the back scarp inclination is flatter, and the boundary transition is smoother, the coarse mesh results are much closer to those obtained using the fine mesh model (Case 3).

Limitations in the CLARA model make it impossible to study the effects of the mesh (or column) size on the stability analysis results for three-dimensional models. In our opinion, the effects should be similar to those observed for the two-dimensional models because the issues discussed above would also apply to the three-dimensional geometry. In fact, we would expect that in the three-dimensional cases this discrepancy would be even greater because of the larger percentage of columns involved at the boundary of the two materials near the base of the slide surface and the increased steepness of the side slopes when compared to the inclination of the back scarp considered in the two-dimensional analyses. Therefore, it should be possible to estimate a three-dimensional factor of safety for a fine mesh model, by increasing the three-dimensional results for the coarse mesh model in the same proportion as the increase calculated from the coarse mesh to the fine mesh results for the two-dimensional model. This was done in the last column of Table 4, which contains the estimated three-dimensional fine mesh factors of safety.

Summary Of Three-Dimensional Stability Analysis

For the standard CLARA analyses, the computed increase in factor of safety due to three-dimensional effects was on the order of 11 to 12 percent, when compared to the two-dimensional (fine mesh) CLARA factors of safety for the same case. In these analyses, the horizontal stresses acting on the sides of the three-dimensional surface were not considered in calculating the resistance to sliding, and the three-dimensional mesh was limited to coarse spacing.

For the "refined CLARA" analyses, the computed increase in factor of safety due to three-dimensional effects was on the order of 30 to 35 percent, when compared to the two-dimensional (fine mesh) CLARA factors of safety for the same case. These increases correspond to the average of the factors of safety computed for scenarios C and D, considered to be the more appropriate scenarios for incorporating the effects of horizontal stresses on the sides of the sliding surface, and include a correction for mesh coarseness.

In our opinion, the results from the refined CLARA analysis provide the best estimates of the three-dimensional factors of safety for the truncated ellipsoidal surfaces.

SUMMARY

This paper has presented the results of two-dimensional dynamic deformation analyses and three-dimensional post-earthquake stability analyses that were completed in support of the design of dam safety modifications for Casitas Dam. The deformation analyses indicated that significant deformations would be expected in the upstream section of the embankment dam, but that freeboard would be maintained at the downstream edge of the crest of the modified dam. The magnitudes of the calculated deformations in the upstream section of the dam are highly dependent on the liquefied strength of the foundation soils. However, it is likely that the deformations will be sufficiently large that transverse cracks will develop in the embankment, and the modifications include a defensive filter layer to address the transverse cracking. The three dimensional stability analyses indicate three-dimensional factors of safety for the upstream slope that are between 10% and 35% higher than corresponding two dimensional factors of safety, depending on the particular strength estimates and sliding surfaces used in the model. Based on these results, the two-dimensional stability analyses may be over-estimating the deformations in the upstream section of the dam.

ACKNOWLEDGEMENT

The authors would like to acknowledge the contributions of Drs Yoshi Moriwaki and Phalkun Tan, of the URS Greiner Woodward Clyde Santa Ana, California office, who provided technical guidance and review of the analyses presented in this paper.

REFERENCES

1. Hungr, O. (1988). *CLARA: Slope Stability Analysis in Two or Three Dimensions.* User's Manual. O. Hungr Geotechnical Research, Inc., Vancouver, B.C., Canada.

2. Itasca Consulting Group (1995). FLAC - *Fast Lagrangian Analysis of Continua.* User's Manual.

3. Stark, T., and Eid, H. (1998). "Performance of Three-Dimensional Slope Stability Methods in Practice." *Journal of Geotechnical and Geoenvironmental Engineering*, ASCE, 124(11), 1049-1060.

4. Wright, S. G, Kulhawy, F. H., and Duncan, J. M. (1973). "Accuracy of Equilibrium Slope Stability Analysis." *Journal of the Soil Mechanics and Foundations Division*, ASCE, October, 783-791.

TABLE 1 – MATERIAL PROPERTIES USED IN FLAC DEFORMATION ANALYSES

Material Number	Material Type	Total Unit Weight (pcf)	Effective Cohesion (psf)	Effective Friction Angle	K_{2max}	Minimum G_{max} (ksf)
1	water	62.4	-	-	-	-
2	waste fill	120	0	27	-	3000
3	zone 4	120	0	35	80	4000
4	zone 3	130	0	38	100	6000
5	zone 1 & 2	133	288	32	65	5000
6	zone 1 & 2	133	0	32	65	5000
7	zone 3	130	0	38	100	6000
8	zone 5	135	144	30	65	5000
9	zone 3	130	0	38	100	6000
10	downstream upper alluvium (liquefied)	115	400 to 800 [1], 0 [2]	0 [1], 28 [2]	70	1500
11	downstream upper alluvium (non-liquefiable)	115	0	28	70	1500
12	upstream upper alluvium (liquefiable)	115	400 to 800 [1], 0 [2]	0 [1], 28 [2]	70	1500
13	upstream upper alluvium (liquefiable)	115	400 to 800 [1], 0 [2]	0 [1], 28 [2]	70	1500
14	lower alluvium	125	0	28	90	3000
15	lower alluvium	125	0	28	90	3000
16	lower alluvium	125	0	28	90	3000
17	lower alluvium	125	0	28	90	3000
18	bed rock	150	-	-	-	42000

Note: (1) Residual Shear Strength
(2) Shear Strength Prior to Liquefaction

TABLE 2 – SUMMARY OF CALCULATED TOTAL DEFORMATIONS OF THE CREST

Analysis Case	Horizontal Deformation [1] (feet)		Vertical Deformation [2] (feet)		Remaining Freeboard (feet)
	Upstream Edge	Downstream Edge	Upstream Edge	Downstream Edge	
Initial Modification Cross-Section:					
Liquefied strength = 800 psf	-2	1	-6	-4	14
Liquefied strength = 600 psf	-19	0.6	-25	-5.5	12.5
Liquefied strength = 400 psf	-38	-3.5	-51	-11.5	6.5
Second Modification Cross-Section:					
Liquefied strength = 400 psf	36	2	-50	-3.3	14.7

Notes:
  (1) Horizontal movement definitions: negative is upstream and positive is downstream.
  (2) Vertical movement definitions: negative is downward (i.e., settlement) and positive is upward.

TABLE 3 – SUMMARY OF FREEBOARD VOLUMES[1]

Analysis Case	Initial Freeboard Volume[2] per Unit Length of Dam (ft^3)	After-Deformation Freeboard Volume per Unit Length of Dam (ft^3)
Initial Modification Cross-Section:		
Liquefied strength = 800 psf	2070	205
Liquefied strength = 600 psf	2070	563
Liquefied strength = 400 psf	2070	205
Second Modification Cross-Section:		
Liquefied strength = 400 psf	2430	763

Note:
  (1) Freeboard volume is the total volume of embankment above the reservoir level, elevation 567.
  (2) For comparison, the freeboard volume for the existing dam is 1530 ft^3/ft.

TABLE 4 – SUMMARY OF RESULTS FOR THREE-DIMENSIONAL SLOPE STABILITY ANALYSES

Case	2-D FS Fine Mesh	% Increase in 2-D for Fine vs. Coarse Mesh	3-D Standard FS Coarse Mesh	3-D Refined FS Coarse Mesh	3-D Refined FS Fine Mesh
1	0.99	12	1.10	1.2	1.3
2	1.06	10	1.19	1.31	1.4
3	1.14	2	1.27	1.46	1.5

Note: Reported refined 3-D factors of safety are an average of calculated values for Scenarios C and D.

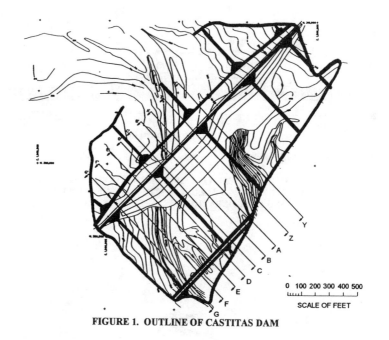

FIGURE 1. OUTLINE OF CASTITAS DAM

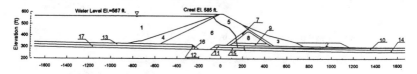

Note: Material properties are listed by number in Table 1

Figure 2a. Cross Section for Existing Conditions

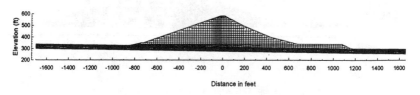

Figure 2b. Finite Difference Mesh for Existing Conditions

**FIGURE 2. DYNAMIC DEFORMATION MODEL
FOR THE EXISTING CONDITIONS SECTION**

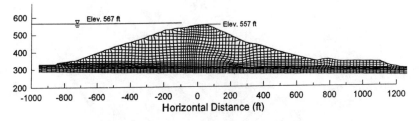

Note: Calculation stopped at 10.5 seconds due to excessive distortion of the mesh

**FIGURE 3. DEFORMED MESH FOR THE EXISTING CONDITIONS SECTION
WITH LIQUEFIED ALLUVIAL STRENGTH OF 600 PSF
(NO EXAGGERATION OF DISPLACEMENTS)**

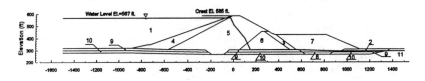

Note: Material properties are listed by number in Table 1

Figure 4a. Cross Section for Initial Modifications

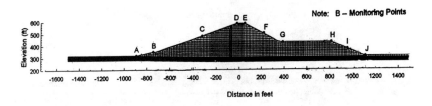

Figure 4b. Finite Difference Mesh for Initial Modifications

**FIGURE 4. DYNAMIC DEFORMATION MODEL
FOR THE INITIAL MODIFICATION SECTION**

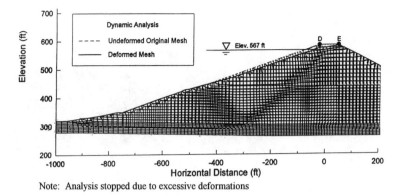

Note: Analysis stopped due to excessive deformations

Figure 5a. Deformed Mesh for a Liquefied Upper Alluvial Strength of 400 psf

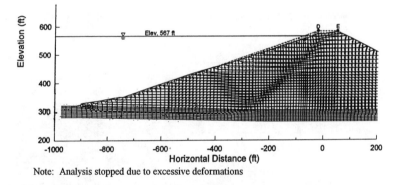

Note: Analysis stopped due to excessive deformations

Figure 5b. Deformed Mesh for a Liquefied Upper Alluvial Strength of 600 psf

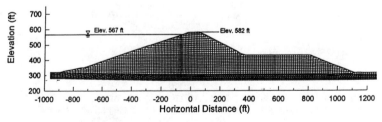

Figure 5c. Deformed Mesh for a Liquefied Upper Alluvial Strength of 800 psf

FIGURE 5. DEFORMED MESH AFTER THE DYNAMIC ANALYSIS FOR THE INITIAL MODIFICATION SECTION (NO EXAGGERATION OF DISPLACEMENTS)

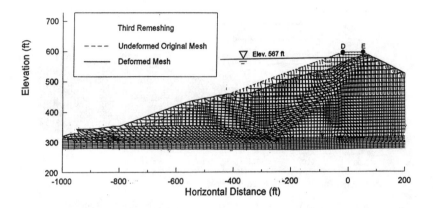

Note: Final deformations achieved after the third remeshing

Figure 6a. Final Deformed Mesh for a Liquefied Upper Alluvial Strength of 400 psf

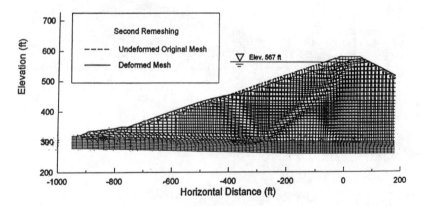

Note: Final deformations achieved after the second remeshing

Figure 6b. Final Deformed Mesh for a Liquefied Upper Alluvial Strength of 600 psf

**FIGURE 6. FINAL DEFORMED MESH FOR THE INITIAL MODIFICATION SECTION
WITH LIQUEFIED ALLUVIAL STRENGTHS OF 400 PSF AND 600 PSF
(NO EXAGGERATION OF DISPLACEMENTS)**

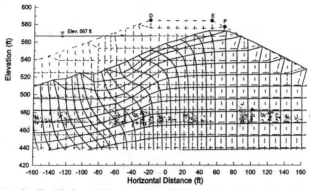

Figure 7a. Final Deformed Crest for a Liquefied Upper Alluvial Strength of 400 psf

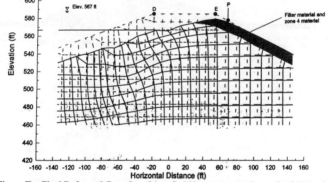

Figure 7b. Final Deformed Crest for a Liquefied Upper Alluvial Strength of 600 psf

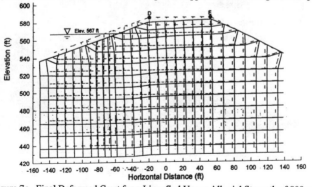

Figure 7c. Final Deformed Crest for a Liquefied Upper Alluvial Strength of 800 psf

**FIGURE 7. FINAL DEFORMATION AT THE CREST FOR THE
INITIAL MODIFICATION SECTION (NO EXAGGERATION OF DISPLACEMENTS)**

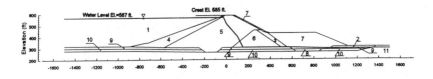

Note: Material properties are listed by number in Table 1

Figure 8a. Cross Section for Second Modification

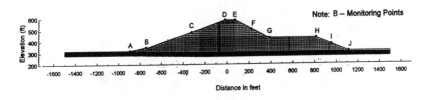

Figure 8b. Finite Difference Mesh for Second Modification

**FIGURE 8. DYNAMIC DEFORMATION MODEL FOR
THE SECOND MODIFICATION SECTION**

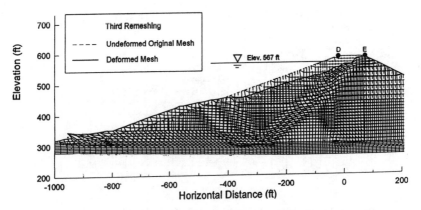

**FIGURE 9. FINAL DEFORMED MESH FOR THE SECOND MODIFICATION
SECTION WITH A LIQUEFIED ALLUVIAL STRENGTH OF 400 PSF
(NO EXAGGERATION OF DISPLACEMENTS)**

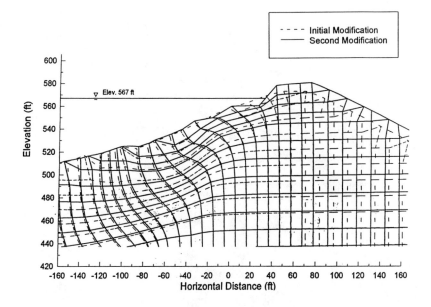

FIGURE 10. COMPARISON OF FINAL DEFORMED MESH FOR INITIAL AND SECOND MODIFICATION SECTIONS WITH A LIQUEFIED ALLUVIAL STRENGTH OF 400 PSF (NO EXAGGERATION OF DISPLACEMENTS)

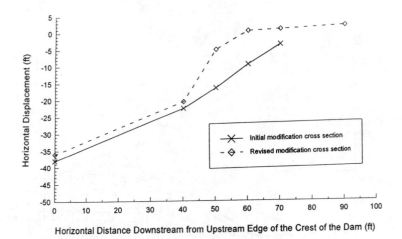

Note: Positive Horizontal Movement is in the Downstream Direction and Negative Horizontal Movement is in the Upstream Direction

Figure 11a. Horizontal Displacement

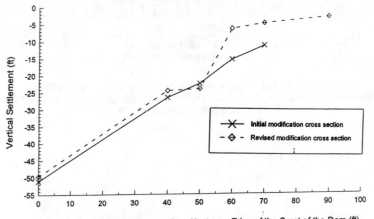

Note: Positive Vertical Movement is Upward and Negative Vertical Movement is Downward

Figure 11b. Vertical Displacement

FIGURE 11. COMPARISON OF HORIZONTAL AND VERTICAL DEFORMATIONS FOR INITIAL AND SECOND MODIFICATION SECTIONS WITH A LIQUEFIED ALLUVIAL STRENGTH OF 400 PSF

MATERIAL PROPERTIES

Number	Material Type	γ_{total} (pcf)	c (psf)	ϕ (deg)
1	Zone 3	125	0	35
2	Zone 1 & 2	130	0	35
3	Zone 5	125	0	35
4	Berm and key fill	125	150	35
5	Upper alluvium (liquefied)	115	600	0
6	Lower alluvium	125	0	28
7	Bedrock	150	5000	45

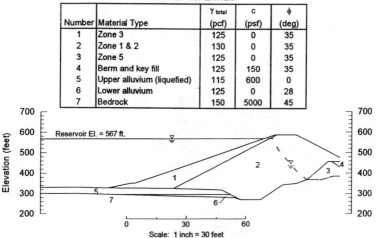

Scale: 1 inch = 30 feet

**FIGURE 12. REPRESENTATIVE CROSS-SECTION FOR
STABILITY ANALYSES – SECTION C**

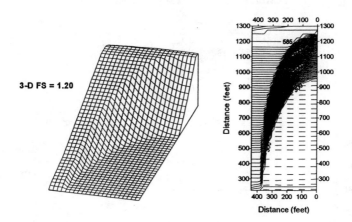

3-D FS = 1.20

FIGURE 13. THREE-DIMENSIONAL STABILITY ANALYSIS RESULTS FOR CASE 2

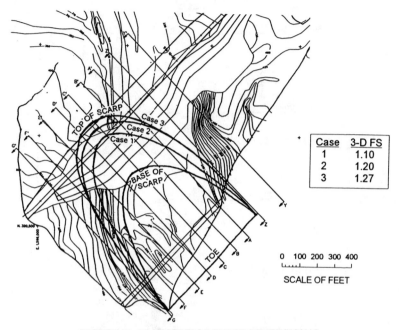

Case	3-D FS
1	1.10
2	1.20
3	1.27

**FIGURE 14. PLAN VIEW OF THREE-DIMENSIONAL
STABILITY ANALYSIS RESULTS FOR CASES 1, 2,3**

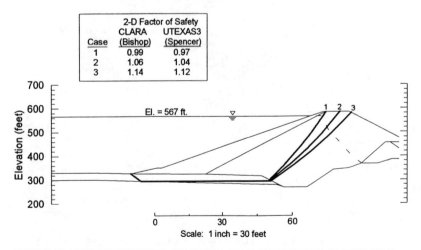

	2-D Factor of Safety	
	CLARA	UTEXAS3
Case	(Bishop)	(Spencer)
1	0.99	0.97
2	1.06	1.04
3	1.14	1.12

FIGURE 15. TWO-DIMENSIONAL STABILITY ANALYSIS RESULTS AT SECTION C

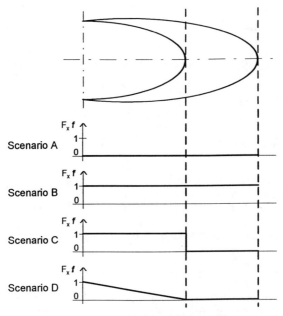

**FIGURE 16. SCENARIOS FOR REFINED
THREE-DIMENSIONAL STABILITY ANALYSIS**

ASSESSING PROBABILISTIC METHODS FOR LIQUEFACTION POTENTIAL EVALUATION

C. Hsein Juang[1], M., ASCE and Tao Jiang[2]

ABSTRACT

In this paper, probabilistic methods developed through logistic regression analyses of field (SPT and CPT) data and those through a newly developed mapping function approach are evaluated and compared. Well-documented cases by U.S. Geological Survey are used in the evaluation and comparison. The results provide a basis for risk-based decisions regarding liquefaction potential.

INTRODUCTION

Simplified methods for evaluating liquefaction potential based on in situ tests, such as Seed et al. (1985) and Robertson and Campanella (1985), are widely used by geotechnical engineers. These methods, referred to herein as *deterministic* methods, have been updated several times (for example, see Youd and Idriss (1997) for SPT-based method, and Robertson and Wride (1998) for CPT-based method). For the convenience of description, the updated CPT-based method by Robertson and Wride (1998) is referred to hereinafter as CPT-RW method. The updated SPT-based method, as presented in Youd and Idriss (1997), is referred to hereinafter as SPT-SI method.

The existing simplified methods, such as CPT-RW and SPT-SI methods, are all developed from some database of field liquefaction performance records at sites that have been characterized with in situ tests. To deal with uncertainties in the data and procedures that were used to develop these simplified methods, it is desirable and necessary to conduct a probabilistic assessment of liquefaction potential that takes into account of these uncertainties.

[1]Professor, Department of Civil Engineering, Clemson University, Clemson, SC 29634-0911. Tel: (864) 656-3322, Fax: (864) 656-2670, E-mail: hsein@clemson.edu.

[2]Research Assistant, Department of Civil Engineering, Clemson University, Clemson, SC 29634-0911.

A number of researchers have contributed to the subject of statistical/probabilistic evaluation of liquefaction. Some have applied probability and statistics to deal with uncertainties that are associated with the simplified methods (for example, Haldar and Tang, 1979; Yegian and Whitman, 1978). Others have conducted logistic regression analyses of field records to establish empirical equations for calculating the probability of liquefaction (for example, Liao et al., 1988; Toprak et al., 1999; Youd and Noble, 1997). The latter approach, first proposed by Liao et al. (1988), has gained significant attention as a tool to characterize the liquefaction boundary curve (the curve that separates liquefaction from no-liquefaction) in the simplified methods.

Toprak et al. (1999) recently performed logistic regression analyses of field records. Their analyses were based on high quality SPT and CPT data compiled by U.S. Geological Survey at sites with and without evidence of liquefaction. Their results show that the CPT-RW boundary curve corresponds to a 50% probability of liquefaction, whereas the SPT-SI boundary curve corresponds to probabilities of 20% to 50%, depending on the magnitude of the corrected blow count for clean sands, $(N_1)_{60,cs}$. The latter results on the SPT-SI boundary curve are generally consistent with those reported by Liao et al. (1988) and Youd and Noble (1997).

Note that the logistic regression approach for calculating the probability of liquefaction is independent of the SPT-SI and the CPT-RW methods. In other words, the equation for liquefaction probability established by the logistic regression has nothing to do with these deterministic methods. They are developed based *solely* on the field data and the form of logistic function. To obtain an understanding of the degree of conservatism of a deterministic simplified method, the boundary curve is often superimposed on the set of probability curves obtained from the logistic regression in a single graph. By observing the position of the boundary curve relative to the probability curves, a statement such as "The SPT-SI boundary curve corresponds to a 20% probability for $(N_1)_{60,cs}$, the clean-sand equivalence of the stress-corrected standard penetration blow count, less than 10 and 50% for $(N_1)_{60,cs}$ greater than 10" can be made (see Toprak et al., 1999). While such statement provides some insight on the conservatism of the deterministic methods, it does not significantly enhance the usefulness of these methods. An engineer who is using, say, the SPT-SI method, will still make his/her decision regarding the liquefaction potential of a soil based on the deterministic result, mostly expressed as a factor of safety. This is because the characterization of the SPT-SI method by the above probability statement does not pinpoint the probability for a specific case that is being evaluated with this simplified method.

Juang et al. (2000) proposed a new approach to calculate the probability of liquefaction based on a mapping function that depends on a particular deterministic method. The mapping function is established to "map" the factor of safety calculated by a particular deterministic method to the probability of liquefaction inferred from field liquefaction performance data. The new approach has been applied to developing a mapping function for the SPT-SI method (Chen and Juang, 2000). In this paper, the mapping function approach is extended to the CPT-RW method, and comparison of the results obtained by the logistic regression and the mapping function approaches is presented.

REVIEW OF DETERMINISTIC METHODS

In the liquefaction evaluation, the cyclic stress ratio may be calculated as (after Seed et al., 1985):

$$CSR_{7.5} = 0.65 \left(\frac{\sigma_v}{\sigma_v'} \right) \left(\frac{a_{max}}{g} \right) (r_d) / MSF \tag{1}$$

where σ_v is the total vertical stress at depth in question, σ_v' is the effective vertical stress at the same depth, a_{max} is the peak horizontal ground surface acceleration, g is the acceleration due to gravity, MSF is the magnitude scaling factor, and r_d is the stress reduction factor. The term MSF is used to adjust the calculated CSR to the reference earthquake magnitude of 7.5. Note that the convention for adjusting the effect of earthquake magnitude is to modify the cyclic resistance ratio (CRR) with MSF. However, it is more logical to include MSF in the calculation of CSR, since both are seismic load parameters, whereas CRR represents soil resistance (Juang et al., 2000).

Idriss (1999) proposed MSF be estimated using the following relationship:

$$MSF = 37.9 \, (M_w)^{-1.81} \quad \text{for } M_w \geq 5.75$$
$$= 1.625 \quad \text{for } M_w < 5.75 \tag{2}$$

where M_w is the moment magnitude of the earthquake.

The term r_d provides an approximate correction for flexibility of the soil profile. In this study, the values of r_d are calculated using the Liao et al. (1988) equation.

The liquefaction resistance of a soil, CRR, may be calculated based on the SPT-SI method (Youd and Idriss, 1997):

$$CRR_{7.5} = \frac{a + cx + ex^2 + gx^3}{1 + bx + dx^2 + fx^3 + hx^4} \tag{3}$$

where a = 0.048, b = −0.1248, c = −0.004721, d = 0.009578, e = 0.0006136, f = −0.0003285, g = −0.00001673, and h = 0.000003741. The variable x in Equation 3 is $(N_1)_{60,cs}$, which is a function of the stress-corrected SPT blow count, $(N_1)_{60}$, and the fines content (Youd and Idriss, 1997). This equation is not a continuous function, and is valid for $(N_1)_{60,cs}$ less than 30. It *must not* be used for $(N_1)_{60,cs}$ greater than 30. The value of CRR approaches infinity as $(N_1)_{60,cs}$ approaches 30 in the SPT-SI method.

In the CPT-RW method, CRR is calculated by the following equation:

$$CRR_{7.5} = 93 \, (q_{c1N,cs}/1000)^3 + 0.08, \quad \text{if } 50 \leq q_{c1N,cs} < 160 \tag{4a}$$
$$CRR_{7.5} = 0.833 \, (q_{c1N,cs}/1000) + 0.05, \quad \text{if } q_{c1N,cs} < 50 \tag{4b}$$

where $q_{c1N,cs}$ is the clean sand equivalence of the stress-corrected cone tip resistance defined by Robertson and Wride (1998). Unlike Equation 3, Equation 4 is a continuous function with no upper bound. However, Robertson and Wride (1998) suggest that the limiting upper value of $q_{c1N,cs}$ is 160. Thus, Equation 4(a) is limited to a maximum $q_{c1N,cs}$ of 160.

PROBABILITY CURVES THROUGH LOGISTIC REGRESSION

The probability equations and data presented in this section are taken from Toprak et al. (1999). For SPT-based logistic regression, 79 data points were used. Among these data, 50 data points were obtained from the 1989 Loma Prieta, California earthquake and 29 data points from other earthquakes including the 1971 San Fernando, the 1979 Imperial, the 1987 Superstition Hills, and the 1994 Northridge earthquakes. The following is the SPT-based probability equation established by Toprak et al. (1999):

$$\ln [P_L/(1-P_L)] = 13.6203 - 0.2820 \, (N_1)_{60,cs} + 5.3265 \ln (CSR_{7.5}) \qquad (5)$$

where P_L is the probability of liquefaction. Note that this equation has been adjusted by the writers to the reference earthquake magnitude of 7.5. Figure 1 shows a set of probability curves defined by Equation 5. It is observed that no liquefied data point falls below the $P_L = 0.3$ curve. It suggests that a design that requires $P_L < 0.30$ may eliminate the occurrence of liquefaction. However, this observation may or may not be valid in cases other than those in the data set analyzed.

Similarly, a probability equation was established by Toprak et al. (1999) based on a logistic regression analysis of a CPT database that consists of 48 data points obtained by the U.S. Geological Survey from the 1989 Loma Prieta earthquake. Their CPT-based probability equation is:

$$\ln [P_L/(1-P_L)] = 12.260 - 0.0567 \, q_{c1N,cs} + 4.0817 \ln (CSR_{7.5}) \qquad (6)$$

Again, this equation has been adjusted by the writers to the reference earthquake magnitude of 7.5. Figure 2 shows a set of probability curves defined by Equation 6. It is observed that no liquefied data point falls below the $P_L = 0.3$ curve. However, this observation may or may not be valid in other cases.

PROBABILITY CURVES THROUGH MAPPING FUNCTIONS

The mapping function approach is first presented by Juang et al. (1999) and later updated in Juang et al. (2000). A mapping function relates the factor of safety (F_S) obtained from a deterministic method such as the SPT-SI method to the probability of liquefaction (P_L) inferred from field liquefaction performance records. Using a *different*

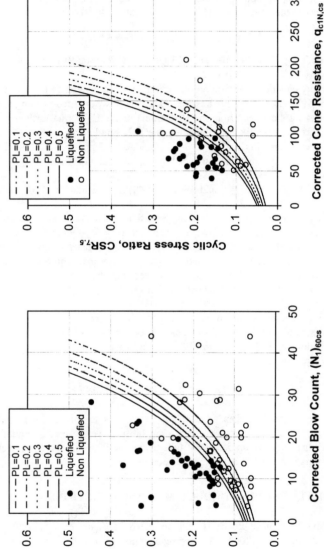

Curves based on Toprak et al (1999)
Data points from Toprak et al. (1999)

Figure 1. SPT-based probability curves

Figure 2. CPT-based probability curves

data set of 233 SPT-based cases, Chen and Juang (2000) established the following equation, referred to herein as the *SPT-SI mapping function*:

$$P_L = 1/[1+(F_S/0.77)^{3.25}] \tag{7}$$

where the factor of safety, F_S, is defined as $F_S = CRR/CSR$, and where CSR is obtained from Equation 1 and CRR from Equation 3. Since CRR is determined from the SPT-SI method (Equation 3), this mapping function is said to have *calibrated* the SPT-SI method based on the field data.

Similarly, the mapping function that calibrates the CPT-RW method (Equation 4) based on CPT field data can be obtained. Using another database of 225 CPT-based cases reported in Juang et al. (2000), the following mapping function is obtained in the present study, which relates F_S determined from the CPT-RW method to P_L inferred from field liquefaction performance records:

$$P_L = 1/[1+(F_S/1.0)^{3.34}] \tag{8}$$

As in Equation 7, the factor of safety F_S in Equation 8 is defined as $F_S = CRR/CSR$. However, CRR is now calculated by Equation 4. CSR is still calculated by Equation 1. Because this mapping function is established based on the premise that CRR is determined by the CPT-RW method (Equation 4), it is referred to herein as the *CPT-RW mapping function*.

SPT-Based Probability Curves

The SPT-based probability curves can readily be plotted based on the SPT-SI mapping function (Equation 7). For example, for $P_L = 0.5$, Equation 7 can be rearranged as:

$$F_S = 0.77 \{(1/P_L) - 1\}^{1/3.25} = 0.77.$$

Since $F_S = CRR/CSR$, where CRR is a function of $(N_1)_{60,cs}$ in the SPT-SI method (Equation 3), it follows that CSR is a function of $(N_1)_{60,cs}$ and a plot of CSR versus $(N_1)_{60,cs}$ can be obtained for $P_L = 0.5$.

Figure 3 shows the SPT-based probability curves. Note that these curves are bounded by $(N_1)_{60,cs} = 30$. If $(N_1)_{60,cs}$ exceeds 30, the *required* CSR level to cause a probability of liquefaction of, say, 0.1 (or any other magnitude), approaches infinity. The implication is that the probability P_L approaches to 0 if $(N_1)_{60,cs} > 30$. This result simply reflects the characteristics of the boundary curve defined in Equation 3. One interesting question is "Is it necessary to have a near-infinity slope at the upper end of the boundary curve?" This is a legitimate question, as the boundary curve in the SPT-SI method was drawn visually based on limited data points. If this restriction is removed (e.g., the SPT-SI boundary curve is allowed to extend beyond $(N_1)_{60,cs} = 30$ without reaching an

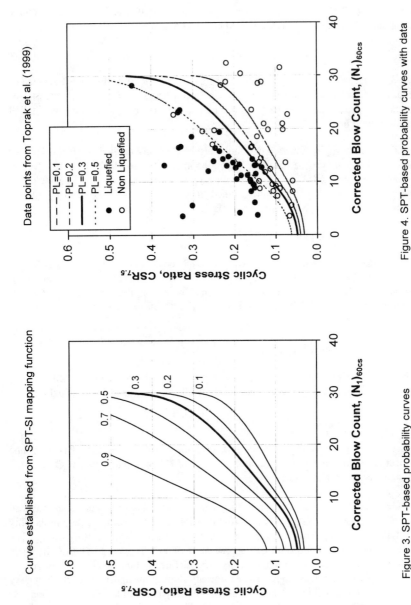

Figure 3. SPT-based probability curves

Figure 4. SPT-based probability curves with data

infinity slope), the probability curves in Figure 3 will likewise extend smoothly without reaching an infinity slope. At present, there are not enough data points to support either proposition. However, allowing the boundary curve to extend without an infinity slope at $(N_1)_{60,cs}$ = 30 yields more conservative results (although not necessarily better results) than does the current SPT-SI method.

It is noted that the P_L = 0.3 curve in Figure 3 coincides with the deterministic SPT-SI boundary curve. This is easily understood from Equation 7, since if P_L = 0.3, then F_S = 1.0, and thus, CSR = CRR. Thus, a plot of CSR versus $(N_1)_{60,cs}$ yields exactly the same curve as the P_L = 0.3 curve shown in Figure 3.

Figure 4 shows a set of the SPT-based probability curves (P_L = 0.1, 0.2, 0.3 and 0.5) along with the SPT field data from Toprak et al. (1999), the same data as those shown in Figure 1. The P_L = 0.3 curve, which coincides with the deterministic SPT-SI boundary curve, bounds liquefied cases.

The observations described above are quite consistent with the results of early studies by Liao et al. (1988), Youd and Noble (1997), and Toprak et al. (1999). However, in these early studies, the deterministic SPT-SI boundary curve was plotted along with a set of probability curves on the same graph. The deterministic SPT-SI boundary curve was *characterized* by probabilities ranging from 20% to 50% based on relative position of the boundary curve to these probability curves. While the result of a logistic regression analysis can be used to characterize a deterministic method such as the SPT-SI method, it does not yield a site-specific probability of liquefaction given a factor of safety calculated from the SPT-SI method. The probability curves obtained through the mapping function approach, shown in Figure 3, on the other hand, are family curves that are specific to the SPT-SI method. Thus, Figure 3 provides a simple way to explore the level of risk of liquefaction based on the factor of safety obtained from the SPT-SI method.

Figures 5 and 6 show a comparison of the SPT-SI probability curves with those obtained from logistic regression by Toprak et al. (1999) for two levels of risk (P_L = 0.3 and 0.5) respectively. The P_L = 0.3 curves are seen to be able to bound liquefied cases, whereas P_L = 0.5 curves do not. Both figures show that at "lower" soil strength, $(N_1)_{60,cs}$ less than about 12 in the case of P_L = 0.3, and less than about 8 in the case of P_L = 0.5, the probability curves obtained from the SPT-SI mapping function are more conservative than the logistic regression curves by Toprak et al. (1999). At "higher" soil strength, however, the trend reverses, i.e., the probability curves obtained from the SPT-SI mapping function are less conservative.

CPT-Based Probability Curves

The CPT-based probability curves can readily be plotted based on Equation 8, the CPT-RW mapping function. Figure 7 shows the CPT-based probability curves obtained from

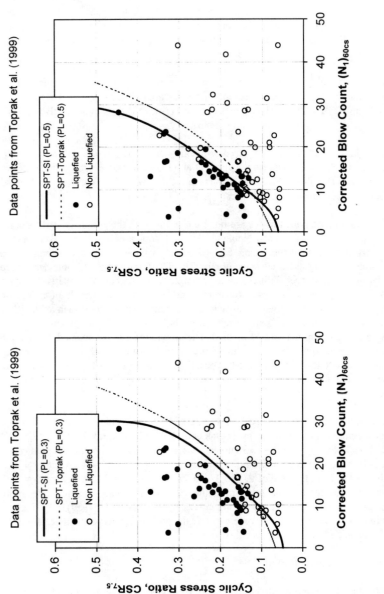

Figure 5. Comparing SPT-based curves at $P_L = 0.3$

Figure 6. Comparing SPT-based curves at $P_L = 0.5$

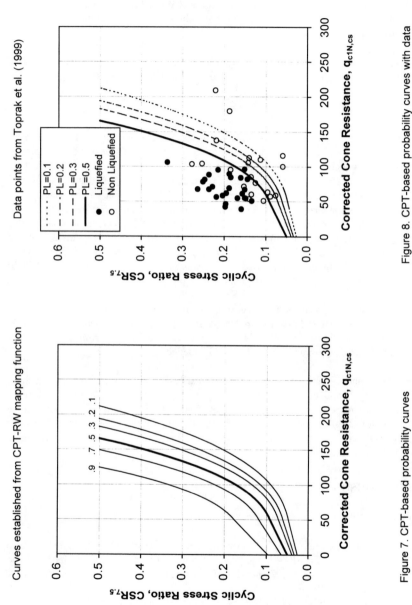

Figure 7. CPT-based probability curves

Figure 8. CPT-based probability curves with data

Equation 8. Note that according to Robertson and Wride (1998), the CPT-RW method is valid for $q_{c1N,cs} < 160$. The CPT-based probability curves shown in Figure 7 reflect this limiting value.

It is noted that the $P_L = 0.5$ curve in Figure 7 coincides with the deterministic CPT-RW boundary curve. This is easily understood from Equation 8, since if $P_L = 0.5$, then

$$F_S = [(1/P_L)-1]^{(1/3.34)} = 1.0$$

and thus, CSR = CRR. A plot of CSR versus $q_{c1N,cs}$ yields the same curve as the $P_L = 0.5$ curve shown in Figure 7. This result is consistent with that obtained by Toprak et al. (1999), which characterized the deterministic CPT-RW boundary curve with a probability of 50% based on the logistic regression analysis. As in the case of SPT study, the characterization of the deterministic CPT-RW boundary curve through the logistic regression analysis does not enable the engineer to determine the probability of liquefaction for a calculated factor of safety. The probability curves obtained in this study based on the mapping function approach, on the other hand, are family curves that are specific to the CPT-RW method. Thus, Figure 7 provides a simple way to explore the level of risk of liquefaction based on the factor of safety obtained from the CPT-RW method.

Figure 8 shows a set of the CPT-based probability curves ($P_L = 0.1, 0.2, 0.3$ and 0.5) along with the CPT field data from Toprak et al. (1999), the same as those shown in Figure 2. The $P_L = 0.5$ curve, which coincides with the deterministic CPT-RW boundary curve, appears to be able to bound liquefied cases.

Figures 9 and 10 show a comparison of the CPT-RW probability curves with those obtained by Toprak et al (1999) through logistic regression for two levels of risk ($P_L = 0.2$ and 0.5) respectively. Both figures show that the CPT-RW curves are slightly more conservative than the logistic regression-based probability curves obtained by Toprak et al. (1999), although the probability curves obtained from both approaches resemble remarkably to each other.

COMPARISON OF SPT- AND CPT-BASED METHODS

First, the following observations are made:

1) The deterministic SPT-SI boundary curve coincides with the $P_L = 0.3$ curve based on the SPT-SI mapping function. This curve distinguishes well the liquefied cases and the non-liquefied cases in the data set compiled by Toprak et al. (1999).
2) The deterministic CPT-RW boundary curve coincides with the $P_L = 0.5$ curve that is based on the CPT-RW mapping function. This curve distinguishes well

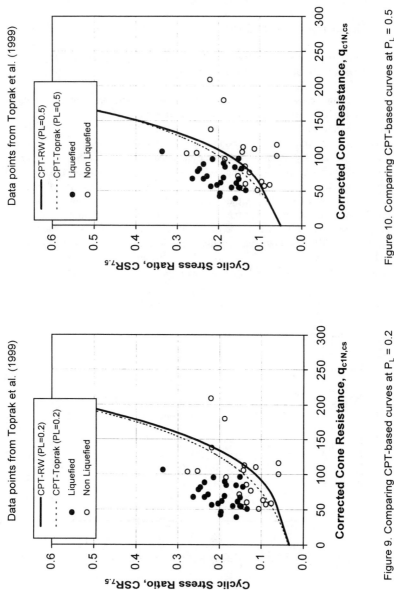

Figure 9. Comparing CPT-based curves at $P_L = 0.2$

Figure 10. Comparing CPT-based curves at $P_L = 0.5$

the liquefied cases and the non-liquefied cases in the data set compiled by Toprak et al. (1999).

3) Characterization of the deterministic boundary curves by logistic regression yield comparable results with those obtained by the mapping function approach. This validates the recently developed mapping function approach by Juang et al. (2000) and Chen and Juang (2000).

In the comparisons presented herein, the objective is to determine which method, the SPT-SI or CPT-RW method, yields a more conservative result. Ideally, if there are sufficient well-documented cases where both SPT and CPT are conducted side by side at the same site at which the soil variability is insignificant, CRR can be calculated using both methods. By plotting CRR values obtained from the SPT-SI method versus those obtained from the CPT-RW methods, a trend may be observed and the more conservative method is identified. Lacking such data, the comparison of the two methods, SPT-SI and CPT-RW methods, may be made using the mapping functions. Figure 11 shows the comparison of the two mapping functions. Recall that the SPT-SI mapping function was developed based on the SPT-SI method using a database of 233 cases, and the CPT-RW mapping function was developed based on the CPT-RW method using a database of 225 cases. Thus, the general characteristics of the deterministic SPT-SI and CPT-RW methods are reflected in the corresponding mapping functions.

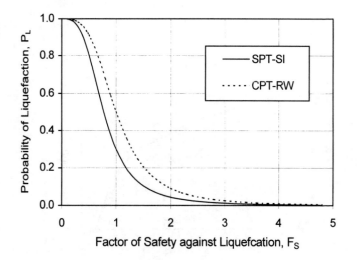

Figure 11. Comparison of SPT-SI and CPT-RW mapping functions

Figure 11 shows that the factor of safety that is required to assure of a risk (probability of liquefaction) of less than a specified value using the deterministic SPT-SI method is lower than that using the CPT-RW method. This indicates that the SPT-SI method is more conservative than the CPT-RW method, since the former requires a smaller factor of safety than does the latter to achieve the same level of risk.

CONCLUDING REMARKS

SPT- and CPT-based probabilistic methods have been assessed in this paper. The CPT-based probability curves obtained from the mapping function approach using the CPT-RW method closely match those obtained through the logistic regression analysis independently performed by other researchers. In the case of the SPT-based probability curves, the agreement between those obtained from the mapping function approach and those from the logistic regression approach is not as good. However, the two sets of probability curves are quite consistent. This result further validates the mapping function approach.

Selection of a proper factor of safety for use with a specific deterministic method, such as SPT-SI or CPT-RW, may be guided by the associated mapping function. Using the associated mapping function, a factor of safety may be selected based on an acceptable level of risk (probability of liquefaction). Thus, a risk-based decision regarding the liquefaction potential may be made. This is one advantage of the mapping function approach over the logistic regression approach.

Using the risk-based comparison, the deterministic SPT-SI method is judged to be more conservative, although not necessarily better, than the CPT-RW method.

ACKNOWLEDGMENTS

This study is supported by the National Science Foundation through Grant No. CMS-9612116. The program director for this grant is Dr. Clifford Astill. This financial support is appreciated. Dr. Ron Andrus is thanked for his review of the manuscript.

REFERENCES

Andrus, R.D., Stokoe, K.H., and Chung, R.M. (1999), *Draft Guidelines for Evaluating Liquefaction Resistance Using Shear Wave Velocity Measurements and Simplified Procedures*, National Institute of Standards and Technology, Report NISTIR 6277, Gaithersburg, MD, 121p.

Chen, C.J. and Juang, C.H. (2000), "Calibration of SPT- and CPT-Based Liquefaction Evaluation Methods," Geotechnical Special Publication (GSP) on Innovations & Applications in Geotechnical Site Characterization, Paul Mayne and Roman Hryciw eds., ASCE, in conjunction with GeoDenver 2000 Conference, Denver, CO., August 2000.

Haldar, A., and Tang, W.H. (1979), "Probabilistic Evaluation of Liquefaction Potential," *Journal of Geotechnical Engineering*, ASCE, Vol. 104, No. 2, pp. 145-162.

Idriss, I.M. (1999), An update of the Seed-Idriss simplified procedure for evaluating liquefaction potential, *Proceedings, TRB Workshop on New Approaches to Liquefaction Analysis*, FHWA-RD-99-165, Federal Highway Administration, Washington, D.C.

Juang, C. H., Rosowsky, D.V. and Tang, W.H. (1999), A reliability-based method for assessing liquefaction potential of sandy soils, *Journal of Geotechnical and Geoenvironmental Engineering*, ASCE, Vol. 125, No. 8, pp. 684-689.

Juang, C.H., Chen, C.J., Rosowsky, D.V., and Tang, W.H. (2000), "A Risk-based Method for Assessing Liquefaction Potential Using CPT," Geotechnique, The Institution of Civil Engineers, U.K., (accepted for publication).

Liao, S.S.C., Veneziano, D., and Whitman, R.V. (1988), Regression model for evaluating liquefaction probability, *Journal of Geotechnical Engineering*, ASCE, Vol. 114, No.4, pp. 389-410.

Robertson, P.K., and Campanella, R.G. (1985), "Liquefaction Potential of Sands Using the CPT," *Journal of Geotechnical Engineering*, ASCE, Vol. 111, No. 3, pp. 298-307.

Robertson, P.K. and Wride, C.E. (1998), Evaluating cyclic liquefaction potential using the cone penetration test, *Canadian Geotechnical Journal*, 35(3), pp. 442-459.

Seed, H.B., Tokimatsu, K., Harder, L.F., and Chung, R. (1985), "Influence of SPT procedures in soil liquefaction resistance evaluations," *Journal of Geotechnical Engineering*, ASCE, Vol. 111, GT 12, pp. 1425-1445.

Toprak, S., Holzer, T.L., Bennett, M.J., and Tinsley, J.C., III (1999), CPT- and SPT-based probabilistic assessment of liquefaction, *Proceedings of Seventh US-Japan Workshop on Earthquake Resistant Design of Lifeline Facilities and Counter-measures Against Liquefaction*, Seattle, August 1999, Multidisciplinary Center for Earthquake Engineering Research, Buffalo, NY.

Yegian, M.K., and Whitman, R.V. (1978), "Risk Analysis of Ground Failure by Liquefaction," *Journal of Geotechnical Engineering*, ASCE, 104(7), pp. 921-938.

Youd, T. L. and Idriss, I. M., eds. (1997), *Proceedings of the NCEER Workshop on Evaluation of Liquefaction Resistance of Soils*, Technical report NCEER-97-0022, T. L. Youd and I. M. Idriss, eds., State University of New York at Buffalo, NY.

Youd, T.L., and Noble, S.K.(1997), "Liquefaction Criteria Based Statistical and Probabilistic Analysis," *Proceedings of the NCEER Workshop on Evaluation of Liquefaction Resistance of Soils*, Technical Report NCEER-97-0022, State University of New York at Buffalo, Buffalo, NY, pp. 201-216.

SEISMIC ZONATION IN THE NEW MADRID SEISMIC ZONE

Salome Romero[1], Student Member ASCE, and Glenn J. Rix[2], Member ASCE

Abstract

Seismic zonation of geologic deposits susceptible to ground motion amplification is implemented by combining in situ measurements, geologic maps, and remote sensing imagery in a geographic information system (GIS). The National Earthquake Hazards Reduction Program (NEHRP) recommends classifying geologic deposits into six profile types based on the average shear wave velocity of the upper 30 meters. Programs such as GIS-HAZUS, created by FEMA, are used to assess seismic risk and estimate potential loss based on the NEHRP classification. Seismic zonation obtained from this project may be incorporated into GIS-HAZUS to improve loss estimates within the New Madrid Seismic Zone (NMSZ).

Conventional seismic zonation entails obtaining field measurements of dynamic properties such as shear wave velocity. This approach requires extensive field testing which may not always be cost efficient, particularly in rural areas. Therefore, an approach for identifying soils susceptible to amplification is presented that merges in situ measurements of dynamic properties with geologic information and available remote sensing imagery. Geologic and soil data were used to identify deposits susceptible to amplification. Remote sensing images were used to delineate alluvial deposits characterized by thick deposits of soft sediments. Geographic information systems (GIS) offer a platform for merging these various data types and assessing the spatial extent of soils susceptible to ground motion amplification.

[1] Graduate Research Assistant, School of Civil and Environmental Engineering, Georgia institute of Technology, Atlanta, GA 30332-0355; phone 404-894-0901; salome.romero@ce.gatech.du
[2] Associate Professor, School of Civil and Environmental Engineering, Georgia Institute of Technology, Atlanta, GA 30332-0355; phone 404-894-2292, glenn.rix@ce.gatech.edu

164 SOIL DYNAMICS AND LIQUEFACTION 2000

Seismic zonation of the New Madrid Seismic Zone (NMSZ) is currently underway in a cooperative effort with the Association of Central United States Earthquake Consortium (CUSEC) State Geologists and the U.S. Geological Survey. Available in situ measurements of shear wave velocity are continuously being collected and archived for this region. The resulting seismic zonation may be used to identify areas requiring additional field testing.

Introduction

The New Madrid Seismic Zone (NMSZ) located in mid-America is a tectonically active region centered along the Mississippi River. The NMSZ extends from western Tennessee and northeast Arkansas to southeastern Missouri. Three earthquakes in the winter of 1811-1812 with estimated magnitudes ranging from 7.3 to 7.8 destroyed the town of New Madrid, Missouri and caused extensive liquefaction of alluvial deposits. Although large magnitude earthquakes such as these are uncommon in mid-America, the region has experienced at least one earthquake with a body-wave magnitude greater or equal to 4.75 every two years (Hays, 1980). Large ground motions have been measured from smaller magnitude earthquakes in the area due to the susceptibility of soft deposits of alluvium and loess to ground motion amplification.

The region does not currently have adequate seismic building codes that consider amplification of ground motions. The older building stock commonly found in mid-America is susceptible to structural failure under seismic loads. Furthermore, an earthquake in mid-America would affect a greater area than an earthquake of comparable magnitude in California due to the lower attenuation in the Eastern United States (Shedlock and Johnston, 1994). Hence, although the NMSZ is considered to have a low seismic hazard compared with California, the seismic risk is considerable due to the geology, lack of adequate seismic building codes, and older building stock. This study focuses on assessing the potential for ground motion amplification in the New Madrid Seismic Zone by analyzing several data sources including geologic and soil data, in situ measurements, and remote sensing images. Based on the seismic zonation of soils susceptible to strong ground motions, subsequent analyses of risk assessment and loss estimation may be performed.

Ground Motion Amplification

Ground motion amplification is responsible for extensive damage in areas underlain by thick deposits of soft sediments. Significant amplification of seismic waves through low shear wave velocity layers caused localized damage in the 1989 Loma Prieta and 1985 Michoacan earthquakes. For example, parts of San Francisco such as the Marina District and the Mission Bay area, located more than 80 kilometers from the epicenter, experienced strong ground shaking due to poorly compacted, loose sediments underlying these areas (Seed et al., 1990). Similarly, in the Michoacan earthquake, areas of Mexico City underlain by soft, thick deposits of old lake-bed sediments experienced stronger shaking than areas underlain by shallow bedrock (Seed et al., 1988).

Ground motion amplification is more pronounced for weak ground motions than for strong ground motions. Therefore, amplification is typically greater for smaller magnitude earthquakes or in areas located at a distance from the epicenter of the event. For example, the largest amplifications in both the Loma Prieta and the Michoacan earthquakes were recorded at significant distances from the epicenter where the amplitude of propagating seismic waves is generally considered to be less due to attenuation.

Two mechanisms contribute to amplification. The first mechanism considers the difference between the frequency of seismic wave energy and the natural frequency of the deposit. A simple estimate of the natural frequency of a geologic deposit (f_n) is given by

$$f_n = \frac{v_s}{4H} \qquad (1)$$

where v_s is the shear wave velocity of the deposit and H is the thickness of the deposit. If the frequency of the seismic wave is approximately equal to the natural frequency of the deposit, amplification will occur increasing the amplitude of ground motion significantly at the natural frequency.

The second mechanism of amplification considers the impedance contrast between two geologic deposits. Impedance (Z) is defined as

$$Z = \rho v_s \qquad (2)$$

where ρ is the mass density of the deposit. Amplification increases as the ratio of impedance between two layers increases. Both mechanisms of amplification are directly related to the shear wave velocity of the deposit. Low shear wave velocity soils, particularly when underlain by hard, crystalline rock, will amplify ground motions causing potential ground or structural failure. Although soil deposits behave as low pass filters by removing high frequency (short period) energy, amplification of longer periods may cause damage to structures with low natural frequencies such as tall buildings and bridges. Therefore, the design of structures in regions characterized by low shear wave velocities needs to consider amplification of ground motions.

Seismic Hazard Mapping of Geologic Deposits. The National Earthquake Hazards Reduction Program (NEHRP) provisions for new structures recommend categorizing the susceptibility to ground motion amplification based on the average shear wave velocity in the upper 30 meters (NEHRP, 1997). NEHRP separates deposits into six site classes as described in Table 1. Maps of expected spectral acceleration assume a soil layer of site class C overlaying bedrock of site class B. Factors accounting for local site effects increase or decrease expected ground motions according to the actual conditions. Risk assessment and loss estimation procedures such as the GIS-based HAZUS program created by the Federal Emergency Management Agency (FEMA) assume site class D as the default (NIBS, 1997). However, although site class D is relatively conservative, it does not consider soils susceptible to collapse under seismic loads such as alluvium or loess and assumes an equal probability of failure. This study distinguishes among the NEHRP recommended soil classes using several data sources including in situ measurements, geologic information, and remote sensing images.

Table 1   NEHRP Site Classification (NEHRP, 1997)

Site class	Description	Ave. Vs in upper 30 m (m/s)
A	Hard rock	$V_s > 1500$
B	Rock	$760 < V_s \leq 1500$
C	Very dense soil and soft rock	$360 < V_s \leq 760$
D	Stiff soil $15 \leq N \leq 50$ or 50 kPa $\leq s_u \leq 100$ kPa	$180 \leq V_s \leq 360$
E	Profile with more than 3 m of soft clay with PI > 20, w $\geq 40\%$, and $s_u < 25$ kPa	$V_s < 180$
F	1. Soils sensitive to failure or collapse under seismic loading such as liquefiable soils, quick and highly sensitive clays, collapsible weakly cemented soils. 2. Peats/highly organic clays with a thickness greater than 3 m. 3. Very high plasticity clays with a thickness greater than 8 m and PI > 75. 4. Very thick soft/medium stiff clays with a thickness greater than 36 m.	

Geology of the NMSZ

The geology of the New Madrid Seismic Zone is potentially susceptible to ground motion amplification due to the presence of deep deposits of soft alluvial and aeolian sediments. The dominant structural feature of the NMSZ is the Mississippi embayment composed of deep, unconsolidated sediments with a maximum thickness of 1000 meters. The Mississippi embayment is a syncline extending from southern Illinois to the Gulf of Mexico centered along the Mississippi River. The embayment dates to the Paleozoic period when crustal movement created a basin for the deposition of alluvial and eroded sediments (Cushing et al., 1964). The soft deposits in the Mississippi embayment are potentially susceptible to ground motion amplification due to their low shear wave velocity. Alluvial deposits are unconsolidated sediments distributed along the floodplains and tributaries of the Mississippi and Ohio Rivers. Loess deposits make up the upland, terrace regions and are aeolian deposits consisting of silt-sized particles characterized by a metastable structure with slight cementation and low density (Bell, 1983). Loess is prevalent in mid-America and formed from the erosion of bedrock and subsequent deposition of sediments. Due to the internal structure, these deposits may be susceptible to collapse particularly when saturated and subjected to seismic loads.

Previous studies of Los Angeles and San Francisco have identified geologic deposits from the recent Quaternary period as the most susceptible to ground motion amplification. Quaternary deposits particularly those from the Holocene epoch are characterized as unconsolidated loose sediments with low shear wave velocities. Tinsley and Fumal (1985) used geologic information to obtain a preliminary seismic hazard map of the Los Angeles area. Surficial information such as grain size and age of deposits was used to delineate areas potentially susceptible to ground motion amplification. Fumal and Tinsley (1985) supplemented geologic information with in situ measurements from standard penetration tests (SPT) to improve seismic hazard mapping in the Los Angeles area. Both studies demonstrated the importance of including geologic information in seismic zonation.

In this study, several data sources were used to delineate soils susceptible to ground motion amplification in mid-America. Conventional seismic hazard mapping requires extensive field testing to determine the spatial variation of shear wave velocity. However, in rural areas such as mid-America where the seismic hazard is low, an extensive testing program is not practical or economical. Therefore, data from available in situ tests was supplemented with geologic information and remote sensing images obtained from the Landsat satellites to improve seismic zonation in mid-America. A study area located northeast of the NMSZ was selected that included portions of southeastern Missouri, southern Illinois, and the Jackson Purchase area in western Kentucky.

Data Sources

In Situ Measurements. In situ measurements of shear wave velocity were collected from various published sources. Only tests directly measuring shear wave velocity were included in this study. Therefore, shear wave velocity obtained from correlations based on cone penetration tests (CPT) or SPT tests were not included. A total of 193 test results were collected from several sources as listed below:

- Nine test sites were obtained from the Illinois States Geological Survey (Bauer, personal communication).
- Casey et al. (1999) performed three seismic cone penetration tests (SCPT) in Memphis, Tennessee.
- Liu et al. (1997) performed three downhole tests in Arkansas, Missouri, and Tennessee.
- Schneider and Mayne (1998a and 1998b) performed 13 SCPT soundings in Tennessee, Arkansas, and Missouri.
- Street et al. (1997) performed 117 seismic refraction surveys in the Jackson Purchase area situated in westernmost Kentucky.
- Street (1999) performed 48 seismic refraction surveys in the greater Memphis area that includes southwestern Tennessee, northeastern Arkansas, and northwestern Mississippi.

The location of each test was mapped on a geographic information system (GIS) using the latitude and longitude of the site and the average shear wave velocity in the upper 30 meters. Figure 1 shows the spatial distribution of the data obtained and the selected study area. A preliminary analysis of the spatial distribution of these measurements was performed using a geostatistical modeling the variation of material properties with separation distance. The

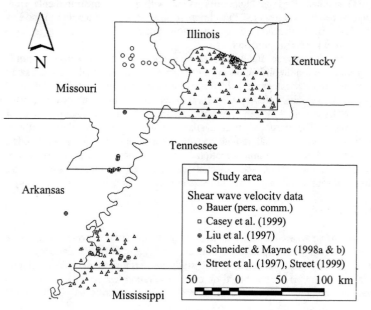

Figure 1 Shear wave velocity measurements in mid-America.

result of kriging was a map of the spatial distribution of average shear wave velocity. The interpolated average shear wave velocity was categorized according to the site classes recommended by the NEHRP provisions. Figure 2 shows the site classes obtained from kriging for the study area.

Based on the kriging analysis, the study area was classified as predominantly site class D. The eastern portion was classified as site class C. Although the map obtained is based on the spatial variation of the average shear wave velocity in the near-surface, there is a large uncertainty in areas with few in situ measurements. Therefore, geologic information and satellite images were introduced to supplement the results from available in situ data.

Geologic and Soil Data. Geologic and soil maps were used to identify Quaternary and particularly Holocene deposits potentially susceptible to ground motion amplification. Geologic maps were obtained from the U.S. Geological Survey and state geological surveys. Geologic maps show structural features and rock units at a typical scale of 1:250,000 or 1:500,000.

Digital soil maps were downloaded from the Natural Resources Conservation Service (NRCS) State Soil Geographic Data Base (STATSGO) as contain information including grain size and index properties such as the liquid limit and the plasticity index of each geologic deposit. The soil maps consider

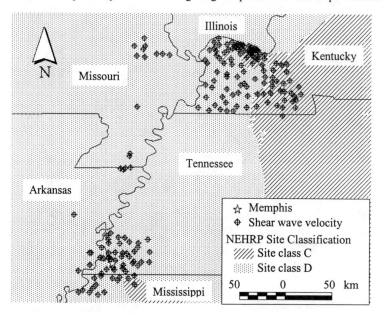

Figure 2 NEHRP site classification obtained from kriging the average shear wave velocity in the upper 30 meters.

only the near-surface soil extending to a maximum depth of 2 meters (6 ft) at a scale of 1:250,000.

Based on the near-surface soil data, clay deposits with an average low mass density ($\rho < 1.4$ g/cm^3), high plasticity index (PI > 20), and high organic content (OC > 5%) in the upper 2 meters were identified as susceptible to ground motion amplification. These soils were classified as site class E based on the 1997 NEHRP provisions. This is a conservative estimate since only the upper 2 meters are considered. However, since these areas are situated along tributaries that consist of loose alluvial deposits, the deposits in the upper 30 meters have a high susceptibility to amplification. Figure 3 shows the deposits identified from the STATSGO soil data.

Remote Sensing Imagery.   Remote sensing systems measure the electromagnetic energy reflected or emitted from the ground surface in various spectral bands ranging from the visible to the microwave regions of the sensing in the visible range. However, other sensors available are capable of measuring energy in longer wavelengths such as the infrared, thermal, and radar (microwave) regions. Analysis of energy in these longer wavelengths enhances

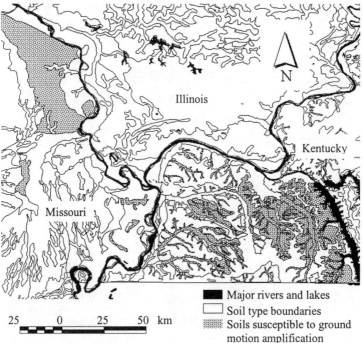

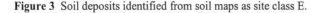

Figure 3 Soil deposits identified from soil maps as site class E.

the discrimination of the ground surface coverage since atmospheric effects such as scattering and absorption are significantly reduced. Two images from the Landsat Thematic Mapper (TM) satellite were acquired from the USGS Earth Resources Observation Systems (EROS) Data Center. The thematic mapper is a second-generation multispectral satellite measuring electromagnetic energy in seven spectral bands as listed in Table 2. The two images obtained cover the selected study area and were geocoded to a standard map coordinate system. The extent of each image is approximately 185 km by 170 km with each pixel representing an area 28.5 m by 28.5 m area. The images were obtained on November 13 and 22, 1986 from the Landsat 5 satellite launched in March 1984. The two images were selected due to the low cloud cover and autumnal acquisition data. Autumnal images were selected due to the lack of vegetation cover allowing imaging of the surficial geology.

Table 2 Landsat TM spectral bands

Spectral Band	Wavelength (μm)	Electromagnetic Region
1	0.45-0.52	Blue-green
2	0.52-0.60	Green
3	0.63-0.69	Red
4	0.76-0.90	Reflected Infrared
5	1.55-1.75	Reflected Infrared
6	10.40-12.50	Thermal Infrared
7	2.08-2.35	Reflected Infrared

Image analysis and classification may be performed using analytical methods or visual interpretation. Several analytical approaches are available which consider the reflectance of a pixel in each spectral band as an element in a feature vector. To minimize redundancy among the available spectral bands, principal component analysis (PCA) may be used to transform the measured spectral bands into linearly independent bands. Classification of pixels is based on minimizing the distance between the feature vector of a given pixel and the class mean. However, analytical methods are computationally intensive. Alternately, preliminary analysis of images may be performed by visual interpretation. Since materials reflect electromagnetic energy differently in each spectral band, surficial coverage may be visually discriminated based on tonal differences.

Sabins (1997) recommends using bands 4, 5, or 7 for analysis in humid climates. Band 4 was selected in this study due to the high contrast in the image. Figure 4 shows the study area in a portion of the two Landsat images.

These figures show the region in band 4. The confluence of the Ohio and Mississippi Rivers is visible in both images. Kentucky Lake is seen at the lower right side of the eastern image (Figure 4b). The city of Paducah, Kentucky is located near the center of the eastern image at the confluence of the Tennessee and Ohio Rivers.

These images were used to delineate alluvial deposits in floodplains and tributaries by identifying features such as oxbow lakes, meander scrolls, and point bars. The floodplains were identified on the imagery based on tonal differences due to soil type and terrain. The boundaries of these alluvial deposits were outlined as shown in Figure 5. The properties of sediments in these features depend on the method of deposition (Kolb and Shockley, 1957). Therefore, based on the depositional characteristics of alluvial sediments, these regions were identified as potentially susceptible to ground motion amplification and classified as site class E.

Integration of Data Sources in a GIS

The information obtained from in situ measurements, geologic and soil maps, and Landsat imagery was combined in a geographic information system (GIS) to obtain a seismic hazard map of the study region. A GIS is a spatial database with powerful analytical capabilities able to handle both raster and vector data types. Raster data is grid-based and includes images and triangulated irregular networks (TIN) such as the output from kriging. Soil maps and in situ measurements are examples of vector data represented by points, lines, and polygons. Vector data such as soil maps and floodplains were converted to a raster format to facilitate integration.

Each data source was combined in a GIS by using the kriged map as the base map or default classification and superimposing the results of the analyses of soil data and remote sensing imagery as shown in Figure 6. The output from kriging was used as the base map since it produced regional coverage of site class. However, an inherent uncertainty exists in these results at significant distances from the location of in situ measurements. To compensate for this uncertainty, the results from analyses of soil data and remote sensing imagery were introduced. Regions delineated by these data sources were classified as potentially susceptible to ground motion amplification. In other words, a preliminary designation of site class E/F may be assumed in these areas. Regions not delineated by these data sources remained unclassified. Therefore, by superimposing the results from the three data sources, an improved seismic zonation of the region is obtained. Subsequent field testing for ground truthing may be performed to assess the potential of strong ground shaking particularly in regions classified as E/F.

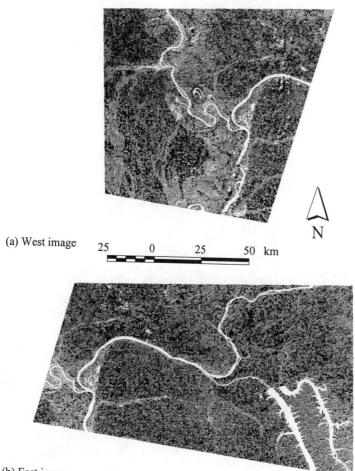

(a) West image   25   0   25   50   km

N

(b) East image

Figure 4 Landsat images of the study area shown in band 4 (0.76-0.90 μm) for (a) west image and (b) east image. The confluence of the Ohio and Mississippi Rivers is seen in both images.

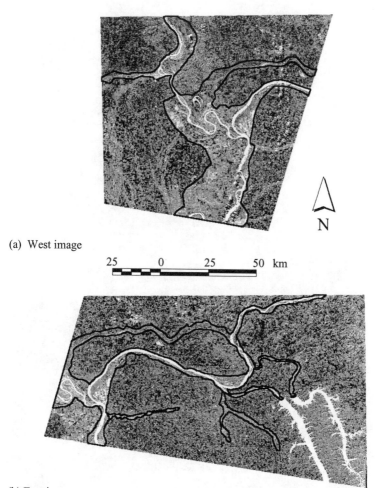

(a) West image

25   0   25   50   km

(b) East image

Figure 5 Floodplains identified from band 4 of the Landsat images for (a) west image and (b) east image.

Conclusions

Seismic hazard mapping is performed for a study area located near the New Madrid Seismic Zone in mid-America. Since extensive in situ measurements of

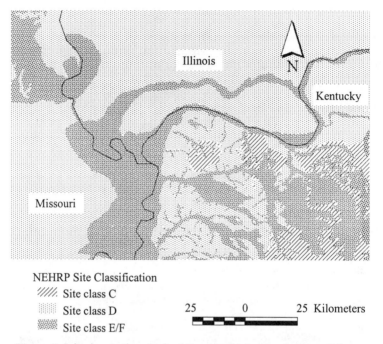

NEHRP Site Classification

//// Site class C

Site class D   25   0   25 Kilometers

Site class E/F

Figure 6 Seismic zonation obtained from combining the analyses of shear wave velocity measurements, soil data, and remote sensing imagery.

shear wave velocity are not available, geologic information and remote sensing images were introduced to supplement in situ data. Based on this study, the default site class for the region was determined to be D. Floodplains along the Mississippi and Ohio Rivers and tributaries were classified as E/F due to the thick layers of soft sediments typically found in these areas. Areas located away from the Ohio and Mississippi Rivers in the upland terrace regions were classified as site class C.

The methodology presented used several data sources to improve seismic zonation by delineating regions based on susceptibility to ground motion amplification. Field testing may be concentrated in these regions to assess the actual shear wave velocity in the upper 30 meters for site specific seismic hazard mapping. Subsequent steps may be taken for seismic retrofitting and disaster planning to reduce risk and loss in these areas.

Acknowledgements

The authors acknowlege the Central United States Earthquake Consortium (CUSEC) State Geologists for providing data. This research is supported by the Mid-America Earthquake Center under National Science Foundation Grant EEC-9701785.

References

Bell, F.G. (1983), *Fundamentals of Engineering Geology*, Butterworths, London, 648 p.

Casey, T., A. McGillivray, and P.W. Mayne (1999), *Results of Seismic Piezocone Penetration Tests Performed in Memphis, Tennessee*, GTRC Project E-20-E87.

Cushing, E.M., E.H. Boswell, and R.L. Hosman (1964), *General Geology of the Mississippi Embayment*, USGS Professional Paper 448-B, 28 p.

Earth System Science Center (ESSC), Penn State University (1999), "Soil Information for Environmental Modeling and Ecosystem Management," Accessed July 1, 1999, http://www.essc.psu.edu/soil_info.

Fumal, T.E. and J.C Tinsley (1985), "Mapping Shear-Wave Velocities of Near-Surface Geologic Materials," *Evaluating Earthquake Hazards in the Los Angeles Region - An Earth-Science Perspective*, USGS Professional Paper 1360, J.I. Ziony, ed., pp. 127-149.

Hays, W.W. (1980), *Procedures for Estimating Earthquake Ground Motions*, USGS Professional Paper 1114, 77 p.

Kolb, C.R. and W.G. Shockley (1957), "Mississippi Valley Geology - Its Engineering Significance," *Journal of Soil Mechanics and Foundations*, ASCE, Vol. 83, No. SM3, pp. 1289-1298.

Liu, H.P., Y. Hu, J. Dorman, T.S. Chang, and J.M. Chiu (1997), "Upper Mississippi Embayment Shallow Seismic Velocities Measured In Situ," *Engineering Geology*, Vol. 46, pp. 313-330.

NIBS HAZUS User's Manual (1997), National Institute of Building Sciences, NIBS Document Number 5200, Washington, D.C.

NEHRP Recommended Provisions for the Development of Seismic Regulations for New Buildings (1997), Prepared by Building Safety Council for Federal

Emergency Management Agency, Washington, D.C., Part 1 - Provisions, 337 p.

Sabins, F.F. (1997), *Remote Sensing: Principles and Interpretation*, 3rd edition, W.H. Freeman & Co., New York, 494 p.

Seed, H.B., S.E. Dickenson, M.F. Rimer, J.D. Bray, N. Sitar, J.K. Mitchell, I.M. Idriss, R.E. Kayen, A. Kropp, L.F. Harder, Jr., and M.S. Power (1990), *Preliminary Report on the Principal Geotechnical Aspects of the October 17, 1989 Loma Prieta Earthquake*, UCB/EERC-90/05, 137 p.

Seed, H.B., M.P. Romo, J.I. Sun, A. Jaime, and J. Lysmer (1988), "The Mexico Earthquake of September 19, 1985 - Relationships between Soil Conditions and Earthquake Ground Motions," *Earthquake Spectra*, Vol. 4, No. 4, pp. 687-729.

Schneider, J.A., and P.W. Mayne (1998a), *Results of Seismic Piezocone and Flat Plate Dilatometer Tests Performed in Blytheville, AR, Steele, MO, and Shelby County, TN*, Interim Report, MAEC Project No. GT-3, GTRC Project No. E-20-677.

Schneider, J.A. and P.W. Mayne (1998b), *Results of Seismic Piezocone Penetration Tests Performed in Memphis, TN and West Memphis, AR*, Interim Report, MAEC Project No. GT-3.

Shedlock, K.M. and A.C. Johnston (1994), "Introduction: Investigations of the New Madrid Seismic Zone," *Investigations of the New Madrid Seismic Zone*, USGS Professional Paper 1538-A, K.M. Shedlock and A.C. Johnston, eds., 6p.

Street, R. (1999), *Shear Wave Velocities of the Post-Paleozoic Sediments in the Memphis, Tennessee Metropolitan Area*, USGS/NEHRP Project No. 1434-HQ-98-GR-0014.

Street, R., E. Woolery, Z. Wang, and I.W. Harik (1997), "Soil Classifications for Estimating Site-Dependent Response Spectra and Seismic Coefficients for Building Code Provisions in Western Kentucky," *Engineering Geology*, Vol. 46, No. 3-4, pp. 331-347.

Tinsley, J.C. and T.E. Fumal (1985), "Mapping Quaternary Sedimentary Deposits for Aereal Variations in Shaking Response," *Evaluating Earthquake Hazards in the Los Angeles Region - An Earth-Science Perspective*, USGS Professional Paper 1360, J.I. Ziony, ed., pp. 101-125.

A PARAMETRIC STUDY ON SEISMIC BEHAVIOR OF
A COMPOSITE DAM

Nien-Yin CHANG[1], Life Member, ASCE, Fatih ONCUL[2]

Abstract

Analyses are performed using a non-linear dynamic analysis code NIKE3D with Ramberg-Osgood non-linear material model and interface algorithm that allows separation and frictional sliding. The earthquake-induced separation along the earth-concrete interface is one of the potential problems in composite dams. A hypothetical 30.48 m (100-ft) high composite dam is analyzed for its behavior under Koyna Earthquake of India, 1967. The concrete and soil slopes on both upstream and downstream sides are varied to observe their effects the dynamic upstream interface behavior. Maximum separation, separation depth, interface nodal accelerations and interface pressure are obtained. The analysis results indicate: 1) Soil-concrete interface can separate under strong shaking, 2) The separation can be significant and the separation along the upstream soil-concrete interface can reach a great depth, 3) Peak crest acceleration can be very high, 4) Maximum dynamic interface earth pressures are significantly higher than static interface earth pressures.

Introduction

Because of the potential for serious consequence of failure, the subject of the seismic safety of earth dams has received immense research attention in the past two and a half decades. The research emphasis on the performance of earth dams under seismic shaking has been on the deformation and loss of free board and the liquefaction of foundation and/or embankment soils. However, a study of the U.S. Bureau of Reclamation online database and NID (National Inventory of Dams) database show that over 40 dams in U.S. are composed of a concrete main dam and earth or rock fill

---

[1] Professor of Civil Engineering, University of Colorado at Denver, email: nchang@carbon.cudenver.edu
[2] Ph.D. Candidate, Civil Engineering Dept., University of Colorado at Denver, email: foncul@ouray.cudenver.edu

Figure 1 *Folsom Dam, California (Height =57.9 m (190 ft)). (Source: USAE Waterways Experiment Station)*

wing dams. A paper titled "The World's Major Dams and Hydroplants (Mermel, 1991) shows 36 concrete-earth fill dams with height over 137.16 m (450 ft) outside U.S. These dams are summarily called "composite dams" in this paper.

Many critical composite dams are located in areas of high seismic risk and their seismic safety is of paramount importance. In a composite dam, the concrete main dam for hydropower generation is extended to the riverbanks by embankment wing dams. As an example; Folsom Dam is shown in Figure 1. The wrap-around sections are the transitional section of a dam where it changes from concrete dam to embankment wing dams. A typical composite section in the wrap-around region has a soil-concrete interface both the upstream and downstream sides and it is illustrated in Figure 2. During strong earthquake shaking, the soil may slip and/or separate (debond) from concrete and, upon the reversal of the direction of motion, the soil may reattach (rebond) itself to the concrete surface. Debonding along the upstream surface allows water to enter the gap created during the process and water is expelled upon rebonding from the gap. The repeated debonding-rebonding can result in a permanent gap due to plastic embankment deformation, internal erosion due to the water pumping action and further the dam failure.

The behavior of the interface during strong earthquake shaking is critical to the seismic safety of a composite dam. Chang (1992) performed dynamic plane strain analyses of a typical composite section of a 57.9 m (190 ft) U.S. dam using FLUSH (Lysmer, 1975), and Oncul (1995) used NIKE2D (Engelmann, 1991) to verify FLUSH findings. Analysis results from the NIKE2D analyses confirm the above mentioned suspicion of the potential for permanent interface separation and the water pumping action.

This paper discusses the effect of cross-sectional geometry of the composite section on the dynamic interface behavior. A 57.9 m (190 ft) high hypothetical dam was selected for this purpose. Due to the high nonlinearity of the problem, the analysis is quite time consuming. The scope of this paper is limited to the investigation of: interface separation potential, separation depth, nodal accelerations, and pressures along the soil-concrete interface.

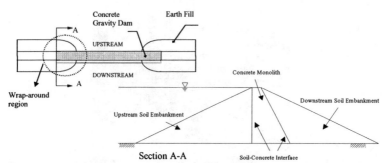

Figure 2 *Plan view of the composite dam and the cross-section with soil-concrete interface.*

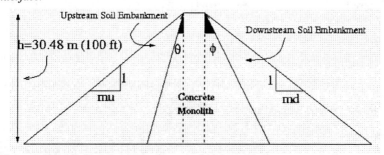

Figure 3 *Hypothetical Composite Dam used in the analyses*

Hypothetical Composite Dam

For this parametric study a cross-section of the 30.48 m (100-ft) high hypothetical dam is selected for analysis as illustrated in Figure 3. The followings are the controlling parameters of cross-sectional geometry: θ, the slope angle of the upstream interface; ϕ, the slope angle of the downstream interface; mu, upstream soil slope; and, md, downstream soil slope as shown in Figure 3.

    The followings are the assumptions considered in the FEM analyses: 1) foundation is assumed as fixed boundary condition; hydrostatic pressure is applied along the upstream soil face; the Coulomb friction prevails along the soil-concrete interface with the coefficient of friction $\mu=0.5$; Ramberg-Osgood non-linear model is used for soil, whereas concrete is assumed as elastic.

Material Properties Used in the Analyses

In the present analysis, the Ramberg-Osgood non-linear model was used to represent soils and linear elastic model for concrete. The equation for Ramberg-Osgood stress-strain relation is given by: $\dfrac{\gamma}{\gamma_y} = \dfrac{\tau}{\tau_y}\left(1 + \alpha\left|\dfrac{\tau}{\tau_y}\right|^{r-1}\right)$

Table 1 *Material Properties*

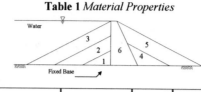

Mat. No	Shear Wave Velocity		τ_y		γ_y (10^{-3})	α	r
	ft/s	m/s	psf	kPa			
1	1500	457.2	1027	49.17	0.11	1.1	2.35
2	1200	365.76	658	31.5	0.11	1.1	2.35
3	900	274.32	370	17.72	0.11	1.1	2.35
4	1500	457.2	1027	49.17	0.11	1.1	2.35
5	1100	335.28	553	26.48	0.11	1.1	2.35
6	Concrete E=2.068E07 kPa (4.32E+08 psf) v=0.20						

where γ=shear strain, τ=shear stress, γ_y=reference shear strain, τ_y=reference shear stress, α=constant$\geq$0, r=constant$\geq$1.

The Ramberg-Osgood relations are inherently one-dimensional, and are assumed to apply to shear components. To generalize this theory to the multi-dimensional case, it is assumed that each component of the deviatoric stress and deviatoric tensorial strain is independently related by the one-dimensional stress-strain equations (Maker, 1995).

Ramberg-Osgood parameters are obtained using the computational procedure proposed by Ueng and Chen (Ueng et.al. 1992). This procedure calculates the Ramberg-Osgood parameters using G_{max} value, and Seed's average modulus and damping ratio versus shear strain curves (Seed, 1970). Table 1 shows all the material properties used in the parametric analysis.

Interface Formulation

NIKE3D uses a penalty formulation for the slidelines with debonding/rebonding and frictional sliding capabilities (Hallquist 1978, 1985). Penetration resistant springs, named penalty "springs" are automatically generated between contact surfaces when the inter-material interpenetration is detected. These springs produce contact forces that are proportional to interpenetration depth. Figure 4 shows the penetration of node "m" into another material where the contact force F_s is given by:

$$F_s = \kappa\delta$$

where κ is the spring constant and δ is the amount of penetration. The constant κ is defined as:

$$\kappa = \frac{f_{SI}K_iA_i^2}{V_i}$$

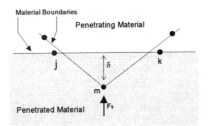

Figure 4 *Penetrating of node "m" into the other material*

where K_i, A_i, and V_i are bulk modulus, area and volume of the penetrated material, respectively. f_{SI} is called penalty scale factor, which allows the user to control the penalty spring stiffness. By choosing relatively stiff penalty springs, the interpenetration can be reduced to insignificant values. Frictional behavior is modeled with Coulomb friction with the coefficient of friction μ taken as 0.5 for the soil-concrete interface.

Table 2 *List of parameters used in the analyses*

CASE 1: θ effect

θ	0	1	2	3	4	5	6	7	8
ϕ	26.6	26.6	26.6	26.6	26.6	26.6	26.6	26.6	26.6
mu	2.0	2.0	2.0	2.0	2.0	2.0	2.0	2.0	2.0
md	2.0	2.0	2.0	2.0	2.0	2.0	2.0	2.0	2.0

CASE 2: ϕ effect

θ	0	0	0	0	0	0	0	0
ϕ	26.6	27.7	28.8	29.9	30.9	32	33	34
mu	2.0	2.0	2.0	2.0	2.0	2.0	2.0	2.0
md	2.0	2.0	2.0	2.0	2.0	2.0	2.0	2.0

CASE 3: mu effect

θ	0	0	0	0	0	0	0	0	0
ϕ	26.6	26.6	26.6	26.6	26.6	26.6	26.6	26.6	26.6
mu	2.0	2.25	2.50	2.75	3.0	3.25	3.5	3.75	4.0
md	2.0	2.0	2.0	2.0	2.0	2.0	2.0	2.0	2.0

CASE 4: md effect

θ	0	0	0	0	0	0	0	0	0
ϕ	26.6	26.6	26.6	26.6	26.6	26.6	26.6	26.6	26.6
mu	2.0	2.0	2.0	2.0	2.0	2.0	2.0	2.0	2.0
md	2.0	2.25	2.50	2.75	3.0	3.25	3.5	3.75	4.0

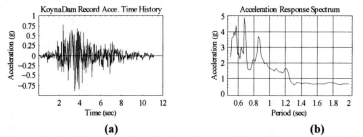

(a)   **(b)**

Figure 5 *(a) Koyna Earthquake Acceleration Time History and (b) Response Spectrum.*

Dam Geometric Parameters

In the initial geometry, the parameters mu, md, θ, and ϕ are assumed to be 2, 2, 0.0 deg., 26.6°, respectively. To analyze their effect on the behavior of composite dam, every parameter is varied, while keeping all others the same as in the initial geometry. Table 2 shows the ranges of parameter values used in the analyses.

Input Motion

In all FEM analyses, the ground motion record of Koyna Dam Earthquake of magnitude 6.5 with, $a_{max}=0.87g$, is used. Figure 5.a shows time history of the ground motion and Figure 5.b gives its response spectrum.

Analyses Results

Each numerical analysis is performed in three stages: gravity turn on, application of hydrostatic pressure, and earthquake loading. Gravity and hydrostatic pressure loading are applied incrementally, whereas dynamic excitation is imposed on a real time basis with a time increment $\Delta t=0.02$ sec.

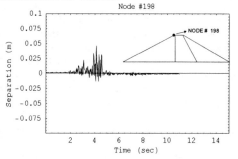

Figure 6 *Separation time history of Node # 198. (Positive means separation)*

After each FEM run, separation, acceleration, and horizontal (σ_x), and vertical (σ_y) stress time histories of all soil and concrete nodes along the upstream interface are gathered. The interface separation takes place at many points along the soil-concrete interface and the separation at each point is repetitive. Figure 6 shows the separation-time history of the Node #198 located on the soil side at the crest of the dam along the upstream interface. It demonstrates the repetitive nature of the separation and closure.

Figure 7 displays three columns of graphs showing the variation of maximum separation, maximum acceleration and interface stresses with the upstream interface angle θ. Static, maximum and minimum dynamic interface stresses are presented. Each row of graphs relates to one θ value, which varies from 0 to 8 degrees at one-degree increment. The graphs on maximum separation show that the soil-concrete interface separation increases toward the crest of the dam with the maximum crest separation of around 0.061 m (0.2 ft), and, in general, it increases with the steepness of the interface with the vertical interface being most critical. The Column 2 of the Figure 7 shows that, at each θ angle, the peak acceleration is amplified as the seismic wave propagates toward the crest of the dam. The acceleration at the crest of the dam achieves a value as large as near 6g, which might be too large. This shows the unrealistic prediction of the crest acceleration during the debonding/rebonding process, which needs further investigation. Excluding the acceleration from the top nodal point, then the maximum nodal point acceleration is about 3g much more realistic.

To further examine the effect of interface angle on maximum separation and the interface earth pressure, the corresponding graphs are collapsed in separate graphs as shown in Figure 8. The maximum separation is found to be around 0.061 m (0.2 feet) and the separation depth can reach as deep as around 18.29 m (60 feet). Figure 8.b shows that the static earth pressure increases with depth with the exception from the depth of 15.24 m to 21.34 m (50 to 70 feet) where a stiffer Type 1 Soil appears as shown in Table 1. The earth pressure is the interface normal stresses between soil and concrete elements along the interface, which is closely approximated by the horizontal normal stress at the concrete nodes or corresponding soil nodes along the interface. Figure 8.c shows that the minimum earth pressure is zero to a depth of around 16.76 m to 19.81 m (55 to 65 ft), which corresponds well with the separation depth shown in Figure 8.a. Figures 8.b, c and d show that minimum dynamic earth pressure is smaller than the static interface pressure and the maximum dynamic interface pressure is much greater than the static interface pressure. The wide band of data variation shows that the effect of upstream interface slope angle θ has significant effects on the magnitude of the interface separation, and static, minimum dynamic and maximum dynamic pressures along the interface.

Figures 9.a to d show the effect of upstream interface slope angle ϕ. The angle ϕ has significant effect on the magnitude of the interface separation as Shown in Figure 9.a. Figure 9.b, c, and d show that the effect of the angle ϕ on interface pressures is not significant.

To examine the influence of upstream embankment slope mu, all analysis results are collapsed as in Figure 10. Figure 10.a shows that the upstream interface slope has a large effect on the depth of separation and the magnitude of the interface separation as indicated in a great separation band. Figure 10.d shows that its effect on the maximum dynamic interface pressure is quite significant. Figures 10.b and c show that the upstream embankment slope has insignificant effect on the static and minimum dynamic interface pressures. Figures 11.a to d show the effect of downstream embankment slope md on the interface separation, and interface pressures on the upstream interface. Its effect on the separation is significant, but its effect on static, minimum earth pressures is insignificant.

Summary and Conclusions

The interface stability of a composite dam under earthquake shaking is quite critical to its seismic safety. Nonlinear finite element analyses are performed to assess the potential for the interface separation, and the influence of the upstream slope angle, upstream embankment slope, downstream interface slope and the downstream embankment slope on four interface performance parameters, magnitude of interface separation, and static, minimum dynamic, and maximum dynamic interface pressures. The conclusions are drawn as follows:

- The interface separation and reattachment are repetitive and are not synchronized at all separation points.
- The upstream interface separation depth reaches a depth of around 16.76 m 19.81 m (55 to 65 ft) from the crest and the separation can be as large as 0.082 m (0.27 ft).
- The upstream interface slope angle imposes significant effect on all four interface performance parameters.
- The upstream embankment slope mu has significant effect on the magnitude of separation and maximum dynamic interface pressure, but has insignificant influence on static and minimum dynamic interface pressures.
- The downstream interface and embankment slopes have significant effect on the upstream separation, but impose insignificant influence on all upstream interface pressures.
- Maximum dynamic interface pressures are significantly larger than the static interface pressure, which is, in turn, larger than the minimum dynamic interface pressures.
- Nodal accelerations are amplified during the upward propagation of seismic wave toward the crest of the dam and the calculated accelerations of all interface soil nodes are reasonable except at the Soil Node #198.

The serious nature of failure along the earth-concrete interface warrants a systematic further study of the subject. The study in progress covers the extension of this 2-Dimensional analysis to the height of the dam of 60.96 m (200 ft), 91.44 m (300 ft), and 121.92 m (400 ft), statistical modeling of the functional relationships of key parameters as performance predictors, 3-D effects, and the effect of material models. The analysis result will then be verified by the centrifugal modeling of the composite dam under seismic shaking.

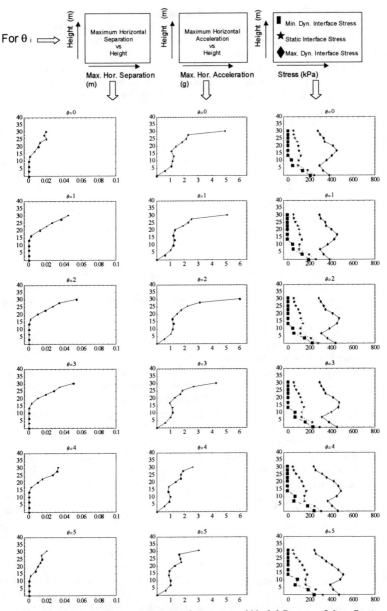

Figure 7 *Max. Separation, Max. Acceleration, and Nodal Stresses (Min., Static, Max.) along the UPSTREAM Interface due to change of θ (in degrees).*

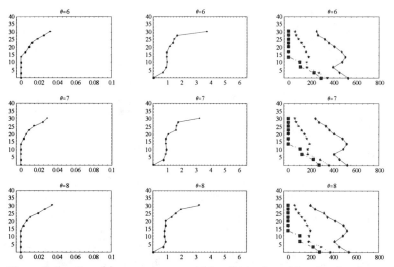

Figure 7 *(continued from previous page) Max. Separation, Max. Acceleration, and Nodal Stresses (Min., Static, Max.) along the UPSTREAM Interface due to change of θ (in degrees).*

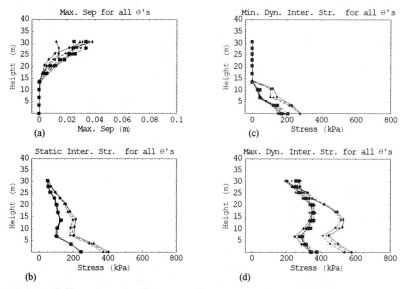

Figure 8 *Collective graphs of Maximum Separation, and interface pressures along the upstream Interface at all θ's.*

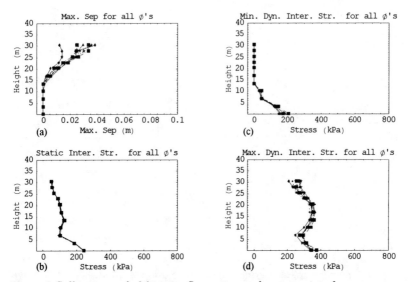

Figure 9 *Collective graphs Maximum Separation, and upstream interface pressures for all ϕ's.*

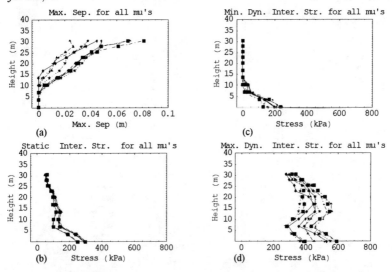

Figure 10 *Collective graphs of Maximum Separation and the upstream interface pressures for all mu's.*

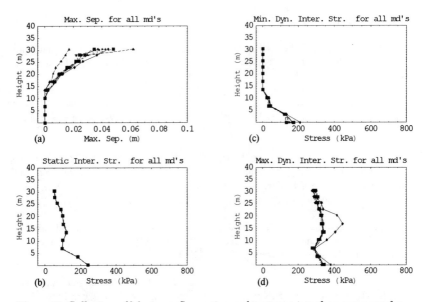

Figure 11 *Collection of Maximum Separation and upstream interface pressures for all md's.*

REFERENCES

Engelmann, B. E. (1991). "NIKE2D: A Nonlinear, Implicit, Two-Dimensional Finite Element Code for Solid Mechanics - User Manual," University of California, Lawrance Livermore National Laboratory, UCRL-MA-105413.

Chang, N. Y., Chiang, H. H. (1992). "Analysis of Possible Separation in A Composite Dam." Unpublished Internal Research Report, University of Colorado at Denver.

Chang, N. Y., Oncul, F., Chiang, H. H., 1997, "Earthquake Induced Separation in a Composite Dam", Eighth International Conference on Soil Dynamics and Earthquake Engineering, SDEE'1997, Istanbul, Turkey, July 20-24, 1997, Volume of Extended Abstracts, A. S. Çakmak M. Erdik, E. Durukal (eds), pp: 264-265.

Hallquist, J. 0. (1978). "A Numerical Treatment of Sliding Interfaces and Impact." Computational Techniques for Interface Problems, AMD Vol. 30, K. C. Park, D.K.Gartling (eds), ASME, New York, pp. 117-133.

Hallquist, J. 0., Goudreau, G. L., Benson, D. J. (1985). "Sliding Interfaces with Contact-Impact in Large-Scale Lagrangian Computations," Computer Methods in Applied Mechanics and Engineering, 51, pp. 107-137.

Lysmer, J., Udaka, T., Tsai, C. F., Seed, H. B. (1975). ``FLUSH: A Computer Program For Approximate 3-D Analysis of Soil-Structure Interaction Problems." Report No. EERC 75-30. Earthquake Engineering Research Center,   University of California, Berkeley, CA.

Mermel,T. W., 1991, "The World's Major Dam and Hydroplants", International Water Power and Dam Construction, June 1991 pp. 67

Oncul, F. 1995, "Earthquake- Induced Separation in a Composite Dam", Master Thesis submitted to University of Colorado at Denver.

Ueng, T., Chen, J. (1992). "Computational Procedures for Determining Parameters in Ramberg-Osgood Elostoplastic Model Based on Modulus and Damping Versus Strain." University of California, Lawrence Livermore National Laboratory, UCRL-ID-111487

Wahl, R. E., Hynes M. E., Yule D. E., Elton, D. J. (1989). "Seismic Stability Evaluation of Folsom Dam and Reservoir Project", U.S Army Waterways Experiment Station, Technical report, GL-87-14 rep. 6, 1989.

In-Situ Liquefaction Evaluation Using a Vibrating Penetrometer

John A. Bonita[1] Student Member ASCE, James K. Mitchell[2] Honorary Member ASCE, Thomas L. Brandon[3] Associate Member ASCE

Abstract

The current approach for using cone penetration test data for liquefaction potential assessment involves the comparison of the seismic stresses imparted into a soil mass during an earthquake to the resistance of the soil to liquefaction. The approach involves an indirect correlation of density and other soil parameters to the liquefaction resistance of the soil mass. A direct approach for evaluating the liquefaction potential of a soil deposit has been proposed through a test using a vibrating piezocone penetrometer. Calibration chamber tests in saturated clean sands indicate that the penetration resistance of a soil mass can be significantly reduced through vibration under certain density and stress conditions. This reduction in penetration resistance is associated with elevated pore water pressures, and is directly related to the liquefaction potential of the soil.

INTRODUCTION

New instrumentation and test procedures have been developed for the direct, in-situ assessment of the liquefaction potential of soil deposits through cone penetration testing. The current approach for using cone penetration test data for the assessment of liquefaction potential involves the comparison of the seismic stresses imparted into a soil mass during an earthquake (CSR) to the resistance of the soil to liquefaction (CRR). The seismic stress imparted into the soil can be determined through the "simplified procedure" established by Seed and Idriss (1971), which is based on the estimated ground accelerations generated by an earthquake, the stress condition present in the soil, and a correction factor accounting for the flexibility of

---

[1] Ph.D. Candidate, Charles E. Via Dept. of Civ. and Env. Engrg., Virginia Tech, Blacksburg VA 24061.

[2] Prof. Emeritus, Charles E. Via Dept. of Civ. and Env. Engrg., Virginia Tech, Blacksburg VA 24061.

[3] Assoc. Prof., Charles E. Via Dept. of Civ. and Env. Engrg., Virginia Tech, Blacksburg VA 24061.

the soil mass (Youd and Idriss 1997). The liquefaction resistance of the soil mass can be obtained through laboratory measures such as static and cyclic triaxial or direct simple shear testing (e.g. Castro 1975; Boulanger and Seed 1995). However, these tests rely on the procurement of high quality samples that are difficult or impossible to obtain in clean sand deposits. Accordingly, for most projects the liquefaction resistance of the soil mass is obtained through field techniques such as standard penetration test (SPT), the static cone penetration test (SCPT), the Becker penetration test (BPT), and shear wave velocity estimations (V_s) obtained through geophysical approaches. All of these field investigative techniques are based on empirical correlations and correction factors, with advantages and disadvantages associated with each method of testing. As an attempt to remove some of the uncertainties associated with these methods, a direct approach for evaluating the liquefaction potential of a soil deposit is proposed through the vibrating piezocone penetrometer test (VCP). The VCP is being developed as part of a research collaboration Virginia Tech and Georgia Tech. Calibration chamber tests are currently being conducted to examine the effects of vibration on the cone penetration resistance and induced pore water pressure. It is hoped that the soil in-situ will respond to the vibration imparted by the cone much in the same fashion as it would from vibrations imposed by an earthquake.

CURRENT EVALUATION APPROACHES

Current methods for evaluating the potential for liquefaction and/or cyclic mobility at a particular site are strongly based on empiricism and correlative statistics. Present state-of-practice procedures for liquefaction evaluation include both laboratory and in-situ test methods that require correction factors that are neither fully understood nor well defined. Laboratory cyclic and simple shear test approaches rely on the testing of high quality soil samples that are often difficult to obtain. Empirical correction factors to account for the initial stress state anisotropy, overburden stress levels, depth effects, cycles to failure, and accumulated strain effects also need to be considered. Similarly, the effects of aging, sample disturbance, and fabric are also difficult to take into account.

Accordingly, for most projects, the general procedure for assessing of the liquefaction potential of a soil deposit involves the use of an in-situ testing technique. To date, the SPT test has been most widely used for this purpose, but unfortunately many corrections must be made to the raw data. These adjustments include, but are not limited to, corrections for hammer efficiency, overburden stress level, fines content, borehole diameter, barrel liner, rod length, and aging (e.g.Youd and Idriss 1997). The corrected N_{spt} value is compared to the capacity of the soil to resist liquefaction (CRR) and the cyclic stress ratio imparted into the soil during an earthquake (CSR). This comparison was initially evaluated in terms of laboratory testing (e.g. Seed and Idriss 1971), but was later expanded using field data through the well known SPT based curves originally proposed by Seed et al. (1984). The cyclic stress ratio (CSR) can also be defined as the average shear stress generated during an earthquake normalized by the effective vertical stress, and is expressed by the following relationship:

$$CSR = \frac{\tau_{avg}}{\sigma_{vo}'} = 0.65 \cdot r_d \cdot \frac{a_{max}}{g} \cdot \frac{\sigma_{vo}}{\sigma_{vo}'} \tag{1}$$

where τ_{avg} is the equivalent shear stress corresponding to 65% of the maximum shear stress, a_{max} is the maximum surface acceleration, r_d is a correction factor accounting for the flexibility of the soil mass, and σ_{vo}/σ_{vo}' is the ratio of the total vertical stress to the effective vertical stress at the depth of interest. The field performance curve based on the SPT test has been normalized to a magnitude 7.5 earthquake, and several methods have been presented by the 1997 NCEER committee for calculating empirical correction factors to account for earthquakes of different magnitude and duration. However, discrepancies exist both in the approach used to determine the correction factor and in the magnitude of the correction factor for a given earthquake, adding additional uncertainty to the liquefaction evaluation process.

More recently, interest has focused more on the use of SCPT for liquefaction potential assessment. The SCPT offers several advantages over the SPT, including a continuous record with depth, the ability to measure pore water pressure generation during penetration, the tests can be performed at proficient speeds and cost, and the test results are amenable to theoretical analysis. Early charts correlating SCPT test results to liquefaction resistance estimations were based on the conversion of N_{SPT} to q_c (e.g. Robertson and Campanella 1985), but the compilation of field data by investigators such as Stark and Olson (1995) and Suzuki et al. (1995) have resulted in the development of SCPT liquefaction curves based directly on SCPT measurements. Similar SCPT correlations were generated by comparing the results from laboratory cyclic loading tests to SCPT values computed from cavity expansion theory (e.g. Mitchell and Tseng 1990). A 1997 workshop sponsored by NCEER was conducted to review developments and to establish professional and academic consensus for evaluating the liquefaction potential of soil deposits. The resulting SCPT correlation recommended by the workshop committee is presented as Figure 1.

The empirical chart in Figure 1 is for a magnitude 7.5 earthquake and level ground conditions, and includes a q_c value that is modified from the measured q_c value through a normalization process, a correction for effective overburden stress, and a correction for fines content in the soil. Much uncertainty was noted by the workshop committee in the correction factors for overburden stress and fines content in the soil. In particular, caution should be used when using the proposed relationship for geological conditions not considered in the original database, such as soils deposited in non-marine environments, reclaimed lands, and residual soils.

The use of the static CPT data for liquefaction evaluation is highly dependent on an indirect correlation of the soil density, as inferred through static penetration resistance, to the estimated cyclic stress ratio imparted into the soil during an earthquake. The method does not directly take into account the deformation behavior of the in-situ soil during cyclic loading, nor does it directly account for the ability of the soil to generate cyclically induced pore water pressures. The method also uses a cyclic stress estimation for the soil that is based on a peak acceleration estimation and only indirectly accounts for frequency and duration of the shaking through the magnitude scaling factors.

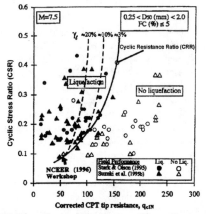

Figure 1. SCPT Liquefaction Evaluation Chart Recommended by the NCEER Committee (after Youd and Idriss 1997)

As noted above, disagreement exists in the academic and professional community on the use of these scaling factors for magnitude correction purposes. As an attempt to remove some of the uncertainty associated with the SCPT technique, a direct approach for evaluating the liquefaction potential of a soil deposit is proposed through using a vibrating piezocone penetrometer test (VCPT). VCPT's have been performed in a large-scale calibration chamber at density and stress conditions that define liquefiable and borderline liquefiable states. The results of the preliminary VCP penetration testing has been presented herein, along with direct comparisons of tests performed under static conditions.

VIBRATING PIEZOCONE PENETROMETER (VCP)

Previous Vibrating Penetrometer Testing
Although both the SPT and SCPT methods summarized above can be useful and reliable in some field conditions, it would be very advantageous to have a test method that provides a direct assessment of the ground's liquefaction resistance. At each test location, specific layers could be evaluated immediately as to their liquefaction potential, regardless of the fines content in the soil and irrespective of the stress state or whether or not the soil is aged, recently deposited, natural, or man made. The vibrating penetrometer is a prospective tool for this purpose, since the intent of the device is to locally induce cyclic pore water pressure in the vicinity of the probe and concurrently measure the penetration resistance during the dynamic loading. Versions of the vibrating penetrometer have been developed and evaluated outside of the U.S. in Japan (Sasaki and Koga 1982; Sasaki et al. 1984; 1986; Teparaksa 1987; Tokimatsu 1988), and in Canada (Moore 1987).

The vibrating penetrometers developed in Japan have been used since the mid 1980's for liquefaction evaluation purposes. All vibrators used in the Japanese testing operate at a frequency of 200 Hz and apply a horizontal centrifugal force ranging from 32 kgf to160 kgf. Each system uses an electric bar type vibrator, similar to that used in concrete placement work, to apply the excitation to the penetrometer. An evaluation of the performance of a 32 kgf vibrating penetrometer was performed by Sasaki and Koga (1982) with different density sands in a steel testing box. The liquefaction potential of the soil was estimated by comparing the penetration resistance of static and vibratory penetration tests and then computing the penetration resistance decrease due to the vibration. Test results indicate that the penetration resistance was reduced by 40% for relative densities less than 50% and that the penetration resistance decrease is less as the effective stress in the soil increases. Similar vibratory cone penetration relationships were also identified by Teparaksa (1987) and Tokimatsu (1988). Teparaksa (1987) also recorded pore pressures on the cone penetrometer at the midface location and noted that the excess pore pressure ratio generated through the vibration, although small in magnitude, was greater than the excess pore pressure ratio generated during the static penetration. This behavior was particularly evident at low confining stresses. Vibratory penetration testing has also been performed in Canada (e.g. Moore 1987). The vibration system used in this field investigation was based on the rotation of an unbalanced mass on top of the cone rods. A vertical motion was induced through this vibration system at average frequencies of 75 Hz, although no estimations of the force applied to the soil was mentioned. A reduction in the penetration resistance measured during the vibration was noted, although no excess pore pressure was identified.

Virginia Tech Vibrating Penetrometer System

Two versions of the VCP have been developed and used to date, with a third unit in the final design stage. The initial prototype vibrator was developed and field tested by the Georgia Tech group as part of this study (Schneider et al.1998; Wise 1998). This prototype unit was mounted behind the cone penetrometer, operated at frequencies between 1-5 Hz, and was based on a pneumatic impulse design. Impulse load tests revealed that the force imparted onto the penetrometer tip through the impact motion was less than 4.4 N. Field tests results reported by Wise (1998) and Schneider et al. (1998) using this unit did not reveal a reduction in vibratory tip resistance when compared to static tests. Excess pore water pressure during vibratory penetration was noted at a few locations, but field variablity in the soil stratum does not allow for the determination of whether the induced pore pressure is attributed to vibratory loading or to normal pore water pressure behavior generation during penetration.

The current version of the VCP consists of a piezocone penetrometer coupled with a rotary turbine vibrator. A schematic of the vibratory unit and associated penetration system is included as Figure 2. The unit is mounted onto the top of the cone penetrometer and generates a vibratory force through the rotation of an eccentrically loaded mass about a horizontally fixed axis. The system is driven by compressed air and operates at a frequency and force that is constant above a minimum input air pressure of 550 kPa (60 psi) and flow rate of 0.4 m³/min (15 cfm). Values of air pressure and flow rate below these minimum values did not generate enough force inside the vibratory unit to turn the eccentrically loaded mass. Rubber dampers were used to isolate the loading system from the calibration chamber. A high-speed data acquisition system was used to collect the vibration at a sampling rate of 750 readings per second. Measurements of the vibration were collected for a 5 second interval during each penetration test when the cone penetrometer reached the center of the sample.

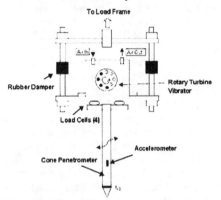

Figure 2. Schematic of Rotary Turbine Vibratory Cone Penetrometer System

Load cells were installed directly below the vibratory unit, and an accelerometer was placed above the cone penetrometer sleeve to measure the magnitude of the force and the frequency of vibration that is both generated by the vibratory unit and imparted into the soil, respectively. An example of the vibratory motion and Fourier transforms obtained from measurements at both the load cell and accelerometer location has been presented as Figure 3.

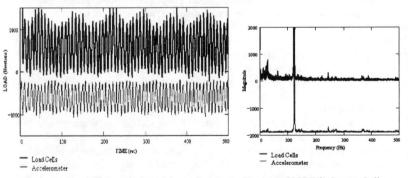

Figure 3. Time History and Frequency Response During a VCPT in Loose Soil

Vibration measurements at the load cell location indicate that the vibratory unit consistently outputs approximately 1300 N (300 lbf) of force at a fairly sinusoidal frequency of 125 Hz. Simultaneous measurements of the frequency of the motion measured at the accelerometer location are very similar to that measured at the load cells. However, the force calculated from the measurements at the accelerometer location is only 850 N (200 lbf), suggesting a 35% loss of the initial generated energy. The loss of energy at the accelerometer location can be attributed to energy dissipation into the soil mass, wave propagation effects as the vibratory stress wave moves through the rod, loose rod and penetrometer connections, and the vertical propagation of energy away into the load frame.

A third vibratory unit is in the final stages of design and evaluation. This unit consists of two eccentrically loaded counter rotating masses that are connected through a rotating gear system. An air motor that allows for different frequencies of motion by adjusting the air-input pressure turns the gears. Different magnitudes of mass can be added to the gears, allowing the vibratory unit to generate a purely vertical vibratory motion that is adjustable in both frequency and force. Large-scale calibration chamber penetration tests are currently being performed in saturated soil pluviated to liquefiable and borderline liquefiable states. Tip resistance (q_c), sleeve friction (f_s), and pore water pressure at the midface (U_1) and shoulder (U_2) locations are continuously measured during each test. The effects of relative density, effective stress level, and vibration force and frequency are currently being evaluated.

V.T. CALIBRATION CHAMBER

Calibration Chamber and Testing Procedure

The Virginia Tech calibration chamber can be described as flexible wall chamber with constant vertical and lateral stress conditions. A schematic of the chamber system has been included as Figure 4. The chamber is 1.5m in diameter, 1.5m tall, and is cylindrical in shape. Sand is placed inside a 40-mil membrane liner inside the chamber by an air pluviator. A perforated plate located at the bottom of the pluviator controls the placement density of the sand. The confining pressure is applied to the sample using air in the annulus between the sample and the cylinder walls. The vertical stress is applied independently of the confining pressure through air pressure bags acting on a free-floating base plate beneath the sample. After the test specimen is fabricated, the saturation procedure is initiated. Carbon dioxide is first percolated through the soil to displace the air in the soil pores, and the sample is then inundated from the bottom to the top. Approximately four to six pore volumes of water are circulated through the sample to displace or dissolve any CO_2 or air entrapped in the void space. The sample is sealed and a water pressure is applied to the interior of the test specimen through an independent water pressure system. B values between 0.92 and 0.96 have been measured using this procedure.

A hydraulic press is then mounted onto the lid of the chamber to push the penetrometer at a constant rate of displacement through a sealed hole in the center of the lid and top plate. Penetration tests are currently being performed both statically and with a vibration applied at the top of the rod. During penetration, the tip resistance, sleeve friction, and pore water pressure are measured and recorded through an automated analog/digital data acquisition system. In the vibratory tests, the force and acceleration generated by the vibrator are also recorded at both the top of the cone rod and at the cone penetrometer.

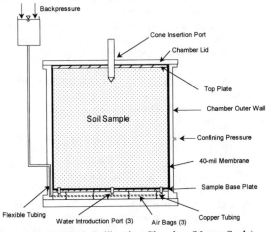

Figure 4. Virginia Tech Calibration Chamber (Not to Scale)

Soil Type and Testing Program

Light Castle sand was used in all of the calibration chamber and laboratory testing as part of this study. The sand has been used extensively by other researchers at Virginia Tech (e.g. Porter 1998). Light Castle sand can be identified as a uniform clean sand (SP) composed of subrounded to subangular quartz grains. A 15 cm^2 Fugro triple element piezocone and a 10 cm^2 Georgia Tech piezocone were used in the testing. Pore pressures are measured at the midface (U_1) and shoulder (U_2) locations on the 15 cm^2 and at the shoulder (U_2) location on the 10 cm^2 cone. Both cones were suggested by their respective manufacturers to be robust in nature capable of withstanding the vibration induced through the vibratory portion of the testing.

Testing in the calibration chamber involved both static (SCPT) and vibratory (VCPT) cone penetrometer tests in loose (D_r = 25%) and medium dense (D_r = 55%) sand samples. Three different effective stress levels and K ~ 0.5 conditions were used in the testing. The three vertical effective stress levels were set at 7.5 kPa, 55 kPa, and 110 kPa to simulate depths of 0.8m, 6.4m, and 12.8m, respectively. Tests were been performed in both dry conditions and saturated states with different levels of backpressure. Measurements of the tip resistance, sleeve friction, and pore water pressure were recorded with depth during the penetration test.

RESULTS OF TESTING

Cone penetrometer tests have been performed in the Virginia Tech calibration chamber under static and vibratory loading, at relative densities of 25% and 55%, under dry and saturated states, and at three different stress levels. Thirty-eight tests have been performed to date to investigate the effects of the vibration on the penetration resistance and pore water pressure and the relationship of the vibratory influence on the liquefaction potential of the soils.

Penetration Resistance

The intent of the vibrating penetrometer device is to locally induce cyclic pore water pressure in the vicinity of the probe and concurrently measure the penetration resistance during the dynamic loading. Preliminary calibration chamber test results indicate that the penetration resistance can be influenced by vibration of the cone penetrometer, provided that the magnitude of the force of vibration is large enough to add a significant dynamic load to the soil. The initial vibrator used for the study was an impact mass unit that imparted less than 0.5 kgf to the soil. Calibration chamber tests using this unit revealed a negligible effect of this dynamic force on the measured penetration resistance. Calibration chamber tests using the rotary turbine vibrator shown in Figure 2 reveal a significant influence of the vibration on the penetration resistance in the loose state at the low and intermediate stress levels. The results of a series of penetration tests performed in a loose soil at the low stress level have been included as Figures 5a and 5b to illustrate this influence. Figure 5a shows the measured penetration resistance value with depth during a series of penetration tests conducted statically and with vibration. All penetration resistance values were corrected for unequal area effects based on a cone area ratio of 0.65 determined in the laboratory. A regression curve was fit through the collection of static and vibratory tests so that a comparison of the penetration resistance values could be made. The regression curves for both the static and vibratory penetration tests have been included as Figure 5b. Also included in Figure 5b is the test data from a combined penetration test where the vibration was started near the midpoint of the sample.

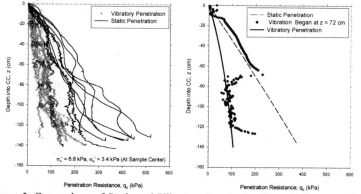

Figure 5. Comparison of Static and Vibratory Penetration Tests in Saturated Loose Samples at Low Stress Conditions

The test results presented in Figure 5 reveal a significant reduction in the penetration resistance at the center of the sample due to the vibration. The center of the sample has been chosen to represent the sample as a whole because is has the lowest influence from the lateral and vertical boundaries. In particular, it appears that the penetration resistance is rapidly reduced over 60% by the vibration for the loose state and low stress conditions. Figures 8a and 8b also reveal that the penetration tests are performed in a reliable and repeatable manner, as indicated by the general agreement of the penetration resistance values for each type test over the small scale presented in the figures (0 kPa to 600 kPa). Further evidence of the quality of the test data is indicated by the close agreement of the combined penetration resistance test data to the regression curves for both the vibratory and static penetration tests. As stated earlier, numerous penetration tests were performed at each stress condition and relative density. Figure 6a are comparative plots of the regression curves obtained from the collection of static penetration tests performed in loose samples at the three stress conditions. Also included in this figure are the regression curves from the collection of vibratory penetration tests performed at each stress state. The penetration resistance value measured during the test (q_c) has been normalized by the vertical effective stress in the sample (σ_v') so that the ratio of (q_c/σ_v') could be compared for all stresses.

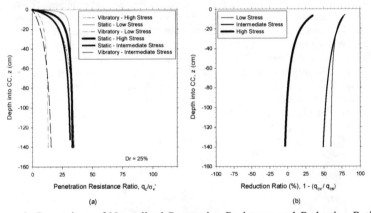

Figure 6. Comparison of Normalized Penetration Resistance and Reduction Ratio with Depth for Tests in Loose Samples at Three Stress Conditions.

It can be seen from Figure 6a that the static penetration tests conducted at the low, intermediate, and high stress conditions normalize to a constant value and that the vibratory penetration tests performed at the low and intermediate stress conditions normalized to a fairly constant value that is significantly less than that of the static tests. It also appears that there was no influence of the vibration on the penetration resistance at high stress levels. Further evidence of the effects of the vibration is noted through Figure 6b, where the reduction ratio, as expressed in a unit of percent, is plotted with depth for tests performed at the low, intermediate, and high stress conditions. Figure 6b indicates that the reduction ratio at the center of the loose sample is about 60% and 65% for the low and intermediate stress conditions, respectively, suggesting that the vibration significantly influenced the penetration resistance. The reduction ratio for this density soil is approximately zero for the high stress state, suggesting that the dynamic load imparted into the soil from the vibration was not significant enough to modify the penetration resistance.

As stated earlier, penetration tests were also performed at all three stress conditions in medium dense (Dr = 55%) samples. Included as Figure 10 is a comparison of the reduction ratio, as measured through the comparison the static and vibratory penetration resistance measured at the center of the sample, versus the vertical effective stress for both the loose and medium dense samples. It can be seen through this figure that there was a significant effect on the penetration resistance at the low stress levels for both low and medium dense states, and that the effect of the vibration decreases as both the effective stress and density of the soil increases. It should also be noted that the reduction ratio never reached a negative value in any of

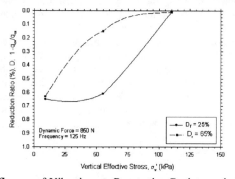

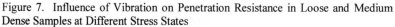

Figure 7. Influence of Vibration on Penetration Resistance in Loose and Medium Dense Samples at Different Stress States

the saturated test performed. Negative values were noted in vibratory tests in dry samples, suggesting that the vibratory penetration resistance was greater than the static. This implies that the vibration imparted into the soil during penetration was densifying the soil ahead of and away from the penetrometer in the dry samples. In the saturated samples, however, excess pore water pressures were initially generated in the soil within the zone of influence of the penetrometer, resulting in a decreased effective stress and subsequent lower penetration resistance.

Pore Water Pressure Measurements

Pore water pressure was measured during each of the penetration tests at either the U_1 and U_2 or U_2 locations depending on whether the 15 cm^2 or 10 cm^2 cone was used. It is generally suggested that the largest pore pressures should be measured at the midface (U_1) location because of the large compressive stresses generated during the penetration (Lunne 1997). This behavior is especially noted in clays, where the permeability of the soils does not allow drainage of the excess pore water pressure. It is also suggested that the smaller, if not negative, induced pore pressures will be measured at the U_2 location, even in loose sands, because of the large shear stresses present along the cone shaft during penetration. Comparisons were made between the excess pore pressure values recorded at the U_1 and U_2 locations in the 15 cm² cone to determine the effect of the element location on the pore pressure value recorded. No difference in the excess pore pressure value was noted. Similarly, no difference in the excess pore pressure values were noted when comparing the measurements from the 15cm² cone to those from 10 cm² cone in identical samples. The level of backpressure in the sample also did not effect the excess pore water pressure recorded.

An example of a comparison of the induced pore pressure recorded during penetration during static and vibratory penetration tests in a loose sample at low stress levels has been included as Figure 8. The excess pore pressure value recorded during these tests has been normalized by the vertical effective stress in the sample and is defined as the pore pressure ratio. It can be seen from this comparison that pore pressure ratio generated during the static penetration tests is essentially zero, suggesting that the permeability of the soil is high enough to allow complete dissipation of any induced pore water pressure generated in the soil during penetration of the penetrometer. A pore pressure ratio of about 0.12 was recorded at the midpoint of the sample (z = 75 cm) during vibratory penetration, suggesting that the pore water pressure was elevated due to the vibration. Similar results were noted for the penetration tests recorded in the loose samples at the intermediate and high stress levels and in the medium dense samples at the low stress levels. The pore pressure ratios measured in the medium dense samples at the intermediate and high stress states were slightly negative for both the static and vibratory tests.

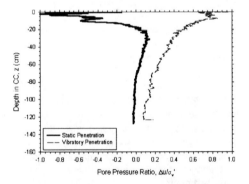

Figure 8. Pore Pressure Ratio Recorded During Penetration in a Loose Sample at Low Stress Levels

CONCLUSIONS

Liquefaction occurs in saturated sandy soils that exhibit the potential for a shear induced volume decrease. This decrease in volume results in a rearrangement of the soil particles into a denser state, resulting in an overall void ratio reduction and expulsion of water from the void space. Soils exhibiting an overall contractive behavior are generally characteristic of low-density material.

The influence of the vibration on the penetration resistance and pore water pressure of loose samples has been shown in Figures 5-8 through a reduction in the measured penetration resistance and elevated water pressures. It has been shown through cavity expansion theory and the works of Salgado et al (1997) that the penetration resistance measured during a cone penetrometer test is controlled by the plastic deformation adjacent to the penetrometer and the elastic deformation at distances away from the penetrometer. It appears that under certain stress and density conditions the dynamic load imparted into the soil by the vibratory unit was significant enough to induce collapse within the soil mass. This most likely resulted in elevated pore water pressures and reduced effective stress conditions both adjacent to the penetrometer and at finite distances laterally and vertically away from the probe.

Significant pore water pressure ratios were not measured during any of the vibratory penetration tests at the U_1 or U_2 locations in the 15 cm^2 cone or at the U_2 location in the 10 cm^2 cone. The vibration of the cone penetrometer is inducing soil collapse both adjacent to and laterally and vertically away from the penetrometer. This collapse of the soil structure results in elevated pore water pressures and liquefaction related conditions. However, the combination of the high permeability of the loose soil, the shear induced densification process associated with the penetration of the penetrometer, the localized measurement zone of the pore pressure transducer, and the set standard penetration rate of the penetrometer (2 cm/sec) does not allow the penetrometer to measure the elevated pore water pressures induced through the vibration. As such, questions have arisen from this investigation on whether or not it is actually possible to measure representative cyclically induced pore water pressures in loose sands during cone penetration testing.

The vibratory penetration tests performed to date included the used of a rotary turbine vibrator that imparted a horizontal component of motion into the soil during the vibration. A vibratory unit is currently being constructed that will impart a purely vertical motion to the soil.

ACKNOWLEDGEMENTS

The research described in this paper was supported by the National Science Foundation and the U.S. Geological Survey. The authors appreciate the generous support provided by this grant. The authors also thank the numerous students and staff that participated in the project.

REFERENCES

Boulanger, R.W., Seed, R.B. (1995). "Liquefaction of Sand Under Bidirectional Monotonic and Cyclic Loading" *J. Geotech. Engrg.*, ASCE, 121(12), 870-878.

Castro, G. (1975)."Liquefaction and Cyclic Mobility of Saturated Sands." *J. Geotech. Engrg.*, ASCE, 101(GT6), 551-569.

Lunne, T., Robertson, P. K., and, and Powell, J. (1997). *Cone Penetration Testing in Geotechnical Practice*, Blackie Academic and Professional.

Mitchell, J.K. and Tseng, D.J. (1990). "Assessment of Liquefaction Potential by Cone Penetration Resistance". *Proc. H.B. Seed Mem. Symp.*, Vol. 2, BiTech Pub. 335-350.

Moore, D. M. (1987). "Evaluation of the Vibrating Cone Penetrometer, and It's Effect on Cone Bearing and Pore Pressure," B.S. Thesis, Univ. of British Columbia.

Porter, J. (1998). "An Examination of the Validity of Steady State Shear Strength Determination Using Isotropically Consolidated Undrained Triaxial Tests," Ph.D. Thesis, Virginia Tech, Blacksburg, VA.

Robertson, P. K., and Campanella, R. G. (1985). "Liquefaction Potential of Sands Using the Cone Penetration Test." *J. Geotech Engrg., ASCE*, 22(3), 298-307.

Salgado, R., Mitchell, J.K., and Jamiolkowski, M. (1997). "Cavity Expansion and Penetration Resistance in Sand." *J. Geotech Engrg., ASCE*, 123(4), 344-354.

Sasaki, Y., and Koga Y. (1982). "Vibratory Cone Penetrometer to Assess the Liquefaction Potential of the Ground." *14th Joint Meeting of the U.S.-Japan Panel on Wind and Seismic Effects*, Washington D.C., 541-555.

Sasaki, Y., Itoh, Y., and , and Shimazu, T. (1984). "A Study of the Relationship Between the Results of Vibratory Cone Penetration Tests and Earthquake Induced Settlement of Embankments." *Proceedings, 19th Annual Meeting of JSSMFE*.

Sasaki, Y., Koga Y., Itoh, Y., Shimazu, T., and, and Kondo, M. (1986). "In Situ Test for Assessing Liquefaction Potential Using Vibratory Cone Penetrometer." *17th Joint Meeting of the U.S.-Japan Panel on Wind and Seismic Effects*, Tsukuba, Japan, 396-409.

Schneider, J. A., Mayne, P.W., and Hendren, T.L. (1998). "Initial Development of an Impulse Piezovibrocone for Liquefaction Evaluation." *International Conference on Physics and Mechanics of Soil Liquefaction*, Johns Hopkins University, Balkema, Rotterdam.

Seed, H. B., and Idriss, I. (1971). "Simplified Procedure for Evaluating Soil Liquefaction Potential." *Journal of the Soil Mechanics and Foundations Division, ASCE*, 97(SM9), 1249-1273.

Seed, H.B., Tokimatsu, K., Harder, L., and Chung, R. (1984). "The influence of SPT Procedures in Soil Liquefaction Resistance Evaluations." *Report No. UBC/EERC-84/15*, Earthquake Engineering Research Center, University of California, Berkley.

Stark, T. D., and Olson, S. M. (1995). "Liquefaction resistance using CPT and field case histories." *J. Geotech Engrg., ASCE*, 121(12), 856-869.

Suzuki, Y., Tokimatsu, K., Taya, Y., and Kubota, Y. (1995). "Correlations Between CPT Data and Dynamic Properties of In-Situ Frozen Samples." *Proc. of the 3rd Int. Conf. on Recent Advances in Geot. Earthquake Engrg and Soil Dynamics*, St. Louis, MO, 1, 249-52.

Teparaksa, W. (1987). "Use and Application of Penetration Tests to Assess Liquefaction Potential of Soils." Ph.D. Thesis, Kyoto University, Kyoto, Japan.

Tokimatsu, K. (1988). "Penetration Tests for Dynamic Problems." *Proceedings of the 1st International Symposium on Penetration Testing, ISOPT-1*, Orlando, FL, AA. Balkema, 117-136.

Wise, C. M. (1998). "Development of Prototype Piezovibrocone Penetrometer for the In-Situ Evaluation of Soil Liquefaction Susceptibility." M.S. Thesis, Georgia Tech, Atlanta, GA.

Youd, T., and Idriss, I. (1997). "Proceedings of the NCEER Workshop on Evaluation of Liquefaction Resistance of Soils." *Report # NCEER-97-0022, NCEER*, Salt Lake City, Utah.

Subject Index

Page number refers to the first page of paper

Author Index

Page number refers to the first page of paper